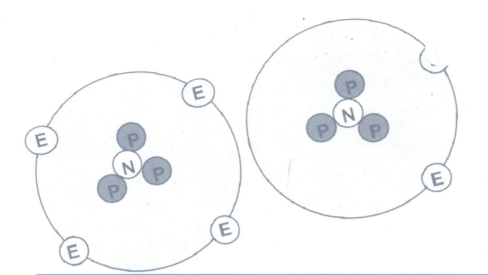

Electrical Theory and Application for HVACR

Turner L. Collins   Earl Delatte   Randy F. Petit, Sr.

esco press

Mount Prospect, IL

www.escogroup.org

Copyright ©2012

VER:2.10116

ESCO Press
PO Box 521
Mount Prospect Il 60056
Phone: 800-726-9696
Fax: 800-546-3726
Website: www.escogroup.org

All rights reserved. Except as permitted under The United States Copyright Act of 1976, no part of this publication may be reproduced or distributed in any form or means, or stored in a database or retrieval system, without the prior written permission of the publisher, ESCO Press.

ISBN 1-930044-32-1

This book was written as a general guide. The authors and publisher have neither liability nor can they be responsible to any person or entity for any misunderstanding, misuse, or misapplication that would cause loss or damage of any kind, including loss of rights, material, or personal injury, alleged to be caused directly or indirectly by the information contained in this book.

Printed in the United States of America

7 6 5 4 3 2 1

Acknowledgements

ESCO Group would like to recognize the following individuals for their contributions to this publication:

Howard "Mell" Greene III, HVAC/R Program, College of Southern Nevada

Rodrigo Jara, Training Department, United Association

Rick Salisbury, Piping Industry Training Center

Table of Contents

Safety and Hazard Awareness — i-iii

- Electrical Shock
- Do not Work Alone
- Electrical Burns
- Non-Conducting Ladders
- Lockout-Tagout Procedures
- Learn First Aid
- Portable Electric Tools
- Safety Guidelines

Chapter 1: What Is Electricity? — 1

- Electrons
- Potential Difference
- Volts
- Measuring volts
- How Electrons Move
- Amperage
- Resistance
- Ohm's Law
- Inductive Reactance
- Capacitive Reactance
- Impedance
- Measuring Resistance
- Wattage

Chapter 2: Circuits and Their Components — 17

- Series Circuit
- Total Resistance in a Series Circuit
- Series Circuit Laws
- Parallel Circuits
- Total Resistance in a Parallel Circuit
- Parallel Circuit Laws
- Combination Series /Parallel Circuits
- Three-Phase Circuits
- Single-Phase Circuits
- The Neutral Wire
- The Safety Ground Wire
- Conductors
- Insulators
- Semi-Conductors
- Circuit Protection
- Loads and Switches
- Transformers
- Low-Voltage Transformers
- Solenoid valves
- Relays
- Sequencer
- Contactors
- Contactor Coil Burn-Out
- Line Starters
- Defrost Timers
- Thermostats
- Heat Anticipators

Chapter 3: Motors   53

- Induction Motors
- Parts of a Motor
- Stator Poles
- Stators
- Rotors
- Motor Speed
- Shaded Pole
- Split-Phase Motors
- Direction of Rotation
- Disconnecting the Start Winding
- CS or CSIR Motors
- PSC Motors
- CSR Motors
- Identifying Hermetic Motor Terminals
- Electronically Communicated Motors (ECM)
- Shaded Pole and PSC Motor Speeds
- Capacitors
- Start Capacitors
- The Run Capacitor
- Capacitor Ratings
- Single-Phase Motor Starting Relays
- Current Relay
- Potential Relay
- Solid State and PTC
- Calculating Motor Horsepower
- Service Factor
- Locked Rotor Amps
- Full Load Amps
- Overload Protectors
- Three-Phase Motors
- Changing the Rotation of a Three-Phase Motor
- Checking Resistance of Windings
- Dual-Voltage Three-Phase Motors
- The Motor Name Plate
- Name Plate Data Definitions
- Variable Frequency Drives
- Variable Speed Drives

Chapter 4: Understanding Wiring Diagrams   79

- Pictorial and Schematic Diagrams
- Pictorials Versus Schematics
- Combining Pictorial & Schematic Drawings
- Line-Side Versus Load-Side
- Ladder Diagrams
- Reading a Wiring Schematic
- Schematic Diagram Symbols

Chapter 5: Automated Control Systems   97

- Why Have Automated Controls
- Building Automation
- System Protocols
- Basic Operating Structure
- Controller Types
- Network Connections

Chapter 6: Troubleshooting   105

- Terms
- Voltmeters
- Lineal Searching
- Split Searching
- Troubleshooting Using a Voltmeter
- Ohmmeters
- Ammeters
- Troubleshooting Switches
- Systematic Troubleshooting
- Voltmeter or Ohmmeter
- Testing Capacitors
- Tips and Suggestions

Glossary   127

Safety and Hazard Awareness

OBJECTIVES:

- Understand the importance of safety
- Understand and practice hazard prevention
- Explain basic safety terms and concepts
- Have a full understanding of lock-out/tag-out procedures

Safety and Hazard Awareness

A safe work environment is the responsibility of everyone involved. Technicians must know how to protect themselves and others when working with electricity.

Always review the wiring diagram, and remove all jewelry before attempting to work on an electric circuit or component.

ELECTRICAL SHOCK

Current is the killing factor in electrical shock. Ohm's Law explains the relationship between voltage, current, and resistance. Human bodies have resistance. When voltage is applied, current flows. If only one-tenth of the current required to operate a ten-watt light bulb were to pass through your chest, the results could be lethal. Most people are killed by 110-volt power, yet we tend to take it for granted.

Ohm's Law states that the amount of current passing through a conductor is directly proportional to voltage applied. If 110 volts were placed across a 500-ohm resistance, the resulting current would be 0.22 amps, or 220 mA. (The *m* stands for milli-, or one thousandth; 1/1,000.) A current of 2 to 3 mA generally causes a tingling sensation. This sensation increases and becomes very painful at about 20 mA. Currents between 20 and 30 mA cause muscle contractions and possibly the inability to let go of the wire. Currents between 30 and 60 mA cause muscular paralysis and difficulty breathing. Breathing at 100 mA current is extremely difficult. Currents between 100 and 200 mA are usually lethal because they cause the heart to go into fibrillation. A 110-V power circuit generally causes between 100 and 200 mA current flow through the bodies of most people.

LOCKOUT-TAGOUT PROCEDURES

An aggressive lockout/tag-out program is one of the best ways to prevent electrical shock. Whenever a piece of equipment is being worked on, it should be disconnected from the power source and locked. The person working on the equipment should carry the only key, to prevent accidental activation. The power supply should be tagged with the following information:
- Name of the person working on the equipment
- What service is being performed
- Reason for service
- Date and time

Some air conditioning units use a contactor with only one set of contacts to disconnect a 240-V circuit. The danger in this is that one line is still supplying power to the unit. If you should happen to touch the live line and ground, a 120-V circuit is completed through your body. Other contactors employ two sets of contacts, disconnecting both of the power lines. This type is safer, but it is always best to disconnect, lock, and tag the power source.

DO NOT WORK ALONE
If you must test a live circuit, have someone with you ready to turn off the power, call for help, or give cardiopulmonary resuscitation (CPR).

LEARN FIRST AID
Anyone working on electrical equipment should take the time to learn CPR and first aid.

ELECTRICAL BURNS
Do not wear rings or other jewelry when working on live electrical circuits. A wristwatch that comes in contact with a live terminal can cause a shock or a severe burn. Never use screwdrivers or other conductive tools in an electrical panel when the power is on. If you were to slip and complete a circuit to ground, a tremendous amount of current would flow, possibly resulting in electrocution. Remember Ohm's Law? If you are working on a 120-V circuit and create a short to ground with a screwdriver blade that has a resistance of 1 ohm, then 120 amps of current will flow. Rings, jewelry, screwdriver blades, etc., generally have resistance values of much less than one ohm.

- ❏ First Aid Kit
- ❏ Jewelry
- ❏ Loose Clothes

PORTABLE ELECTRIC TOOLS
Portable electric tools constructed with a metal frame should have a grounding wire in the power cord. The grounding wire protects the operator from electrical shock if the tool develops a loose connection that could cause the frame to become hot. If this occurs, the grounding wire, not your body, will carry the current to ground, and a breaker or fuse will interrupt the circuit. In some instances, wall receptacles with two prongs may be adapted to a three-prong cord. The adapter has a ground wire that must be connected to a good ground in order for the circuit to provide protection. More modern receptacles have a Ground Fault Circuit Interrupter (GFCI) design. A GFCI receptacle can detect a very small electrical leak to ground, and will open the circuit to prevent further current flow.

It is a good safety practice for service personnel to have a short extension cord with a GFCI receptacle for field use of power tools and equipment.

Many modern handheld power tools are constructed with their wiring and motor housed in a plastic case and are considered double insulated. Battery operated tools use rechargeable batteries, which are both convenient and safe.

NON-CONDUCTING LADDERS

Aluminum ladders can be hazardous if they are accidentally raised into a power line. Non-conducting ladders made of fiberglass or wood should be used. These ladders will help protect you from a shock to ground, but *not* from a shock between two live wires.

Aluminum    **Wood**

Fiberglass    **Extension**

SAFETY GUIDELINES

- When servicing any equipment, unless power is required, shut off the power supply and lock and tag it.
- When working on live circuits, avoid contact with damp or wet surfaces.
- Use only properly grounded power tools.
- Do not wear jewelry.
- Discharge all capacitors before touching the terminals (use a 15K ohm, 2-watt resister).
- Check all test leads and probes for cracks and broken insulation.
- Do not use metal ladders.
- Wear shoes that have insulated heels and soles.
- Always follow the National Electrical Code.
- Always remember that your life and the lives of others depend on your safety practices.

Chapter 1: What Is Electricity?

OBJECTIVES:

- Understand atomic structure and theory
- Identify various electrical terms and laws
- Measure voltage, current, and resistance, and calculate impedance
- Recognize electrical symbols
- Gain knowledge of electrical theory and practical uses

What Is Electricity?

Electricity is a form of energy. Electrical energy is the movement of electrons through a conductor. Other forms of energy include heat, light, chemical (battery), atomic (power plant), and mechanical (motor). Energy cannot be created or destroyed, but it can be converted from one form to another. For example, heat is used to produce electricity, which can be used to produce light. Electrical appliances are designed to convert electrical energy to another form of energy, thereby performing useful work. Some devices produce heat and others produce motion or light.

ATOMS/ELECTRONS

All matter is made of atoms. Atoms are made up of particles called protons, neutrons, and electrons. Protons and neutrons are located at the center, or nucleus, of the atom. Electrons travel in orbits around the nucleus. Protons have a positive charge, electrons have a negative charge, and neutrons have no charge and therefore no effect on the electrical characteristics of the matter. Electrical energy is released when electrons move from one atom to another; electrons can be forced to pass from one atom to another. Atoms try to maintain equal numbers between positive (+) and negative (-) charges (protons vs. electrons). An atom that loses an electron becomes positively charged (+) due to the excess proton. An atom that gains an extra electron becomes negatively charged (-).

The Law of Electric Charges states that like charges repel and opposite charges attract. Excess electrons are attracted to atoms lacking electrons. To perform useful work, a constant and steady movement of electrons must be produced.

POTENTIAL DIFFERENCE

An imbalance of electrons is called a potential difference. A potential difference describes a situation where excess electrons have accumulated, and are waiting for an opportunity to reconnect with atoms that lack electrons. Electrons travel from negative (-) to positive (+) atoms.

There are a variety of methods used to create a potential difference (electromotive force) between two points: friction (static electricity), chemical (battery), thermoelectric (heat), photoelectric (light), and magnetic (generator or alternator).

The presence of a potential difference is sometimes called electromotive force (EMF), which is further abbreviated to "E."

VOLTS

The potential difference (EMF) between two points can be very high or very low. The unit of measurement used to indicate the strength of the EMF is the Volt.

Some typical voltages include: 1.5 Volts for a flashlight cell, 12 Volts for auto batteries, 24 Volts for controls, 120 Volts for homes, and 240 Volts for commercial systems. Voltage can vary from a microvolt (millionths of a Volt) to megavolts (millions of Volts).

The terms potential, electromotive force (EMF), and voltage mean the same thing and can be used interchangeably. Most people refer to EMF as Volts. Remember, electromotive force is NOT electricity. It is the driving force that causes electrons to move from one atom to another.

MEASURING VOLTS

Voltmeters are used to measure the potential difference between two specific points and are available in analog or digital types. Digital meters are much easier to read because they display the voltage directly while analog meters move a pointer across a scale in proportion to the voltage of the circuit and can be easily misread.

All voltage testers, regardless of type, have two probes and the meter indicates potential difference between the two probes. Voltmeters are often used to check electrical power supply. Correct placement of the probes and interpretation of the readings is critical for proper troubleshooting of electrical problems.

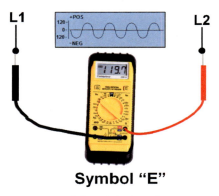

Symbol "E"

Fig. 1-1 (a): Digital Voltmeter   Fig. 1-1 (b): Analog Voltmeter

Electrical appliances are energy-consuming and conversion devices (called loads) and they are designed for connection between a potential difference. A specific voltage must be applied to force electron movement through the device. When testing supply voltage, a maximum variation of plus or minus five percent is generally acceptable. Connecting wires supply the necessary electrons and complete the circuit or pathway for electron flow. When the proper voltage is connected to a load, the load should operate. If the device is supplied the proper voltage and does not operate, it is defective. A voltage tester can quickly reveal this problem.

A voltage tester reads zero when no potential difference exists between the two probes, but will also read zero if voltage and polarity are the same at both probe locations. Additional voltage tests are required to determine whether or not voltage is present.

 Never touch an electrical wire because a zero voltage reading was obtained – you may be reading the same potential (no difference) between the probes!

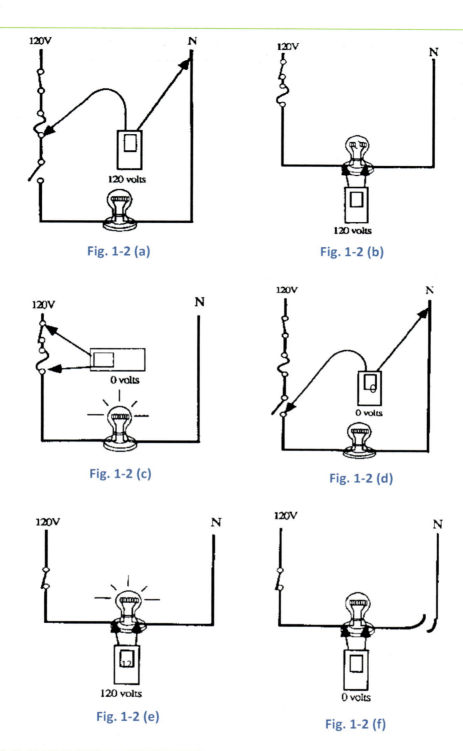

Fig. 1-2 (a)
Fig. 1-2 (b)
Fig. 1-2 (c)
Fig. 1-2 (d)
Fig. 1-2 (e)
Fig. 1-2 (f)

HOW ELECTRONS FLOW THROUGH A CONDUCTOR

Voltage is the force that causes electrons to move, but electrons cannot move unless they have a place to go. Electrical circuits (electrical pathways) are composed of copper or aluminum wires and devices or loads that are designed to control the flow of electrons (current). Power plants produce EMFs, NOT electrons. The electrons are already inside the wires. The EMF produced by the power plant forces free electrons inside the conductor to travel to the next atom within the conductor, much like a domino effect. This electron movement from one atom to another occurs throughout the length of the conductor. The conductors provide the necessary free electrons and provide the proper pathway for electron movement. In a DC circuit (direct current), the electrons travel in one direction only. In an AC circuit (alternating current), the electrons are constantly changing directions. In the United States, power is transmitted at 60 hertz. This means the change from negative to positive occurs 60 times per second.

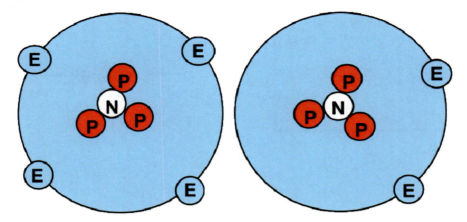

Fig. 1-3: The negative atom has 4 electrons and the positive atom has 2 electrons.

Electricity is often compared to the flow of water. However, water valves and electrical switches operate differently. Water flows when a valve is opened. Electricity flows when a switch is closed; electrons cannot flow through an open switch or a broken wire. Any opening in the circuit (pathway) is much like a drawbridge.

Switches are used in electrical circuits to act as these "drawbridges" for stopping and starting the flow of electrons.

AMPERAGE

The words *ampere*, *amperage*, *amps*, and *current* are frequently used to describe the quantity and intensity of electrons moving through a conductor. Amperage determines how much electricity will be converted to another form of energy. Thus, electrical loads are "energy conversion devices." These energy conversion devices (toasters, light bulbs, motors, etc.) are used to perform useful work.

An ammeter is used to measure the quantity and intensity of electrons flowing inside a wire, or through a load. The letter "I" (for "intensity") is often used to indicate amperage flow. When current flows through a conductor, a magnetic field is created. The clamp-on ammeter is most commonly used on AC circuits and is designed to read the intensity of flow in ONE wire. The magnetic fields will cancel each other out and the ammeter will indicate a reading of zero when clamped around the two wires feeding a load device. If the same wire is coiled through the jaws of the ammeter, the reading will increase by the multiple of each wrap of the coil. With an amperage reading of 20, and the wire coiled 5 times, the actual amperage for the circuit will be 4. Figure 1-4 illustrates a typical clamp-on ammeter. These are available in analog type (needle pointer) or digital readout.

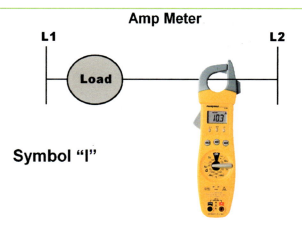

Fig 1-4: The inductive ammeter reads intensity of magnetic field around the wire and converts it to an amperage reading.

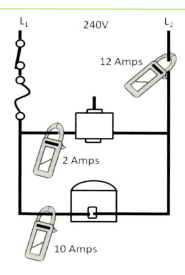

Fig 1-5: Total amperes for a circuit is the sum of the branch circuits.

RESISTANCE

Resistance refers to anything offering opposition to the flow of electrons. It is the resistance that causes energy conversion. Electron flow is energy in motion and must be controlled. The resistance is one factor that controls the amount of electron flow, and thus regulates the rate at which the useful work is performed. A circuit without resistance to control the electron flow is considered shorted.

There are several types of resistance that will be discussed later in this chapter. However, a basic understanding of Ohm's Law is necessary before that discussion.

Fig. 1-6

OHM'S LAW

Ohm's law, discovered by George S. Ohm, defines the exact relationship between voltage (E), amperage (I), and resistance (R). Ohm's Law is used for troubleshooting purposes and designing electrical devices and circuits. The capital letter "R" is often used to indicate resistance. Another symbol for resistance is the Greek letter Ω (omega).

Ohm's Law is best remembered as a pie, as shown in Figure 1-7. To use the pie chart, cover the item to be determined and follow the instructions as indicated by the horizontal or vertical lines. For example, to discover E, you must multiply I by R. To discover I, divide R by E.

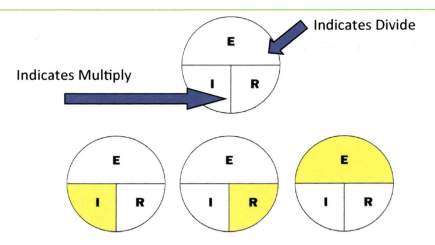

Fig. 1-7: Cover the unknown item and follow the instructions indicated by the horizontal or vertical lines.

RESISTANCE AND REACTANCE

There are three important types of opposition to electron flow: pure resistance, inductive reactance, and capacitive reactance.

PURE RESISTANCE

Pure resistance is opposition to current flow where the current stays in phase with the voltage. Pure resistance can be directly measured with an ohm meter and will only change with temperature. Toasters and electric heaters are examples of loads that have close to pure resistance. (Ohms Law: With a fixed resistance, higher voltage increases amperage and lower voltage decreases amperage.)

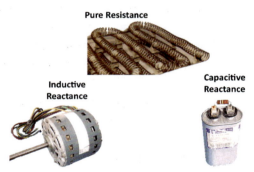

Fig. 1-8: Types of Resistance

INDUCTIVE REACTANCE

When a conductor is wound into a coil, the magnetic lines of force overlap and reinforce each other inducing a counter-EMF or opposing alternating current in the adjacent windings. The counter-EMF is the source of opposition to current flow. A constant direct current has a zero rate-of-change and sees an inductor as a short circuit (it is typically made from a material with a low resistivity). An alternating current has a time-averaged rate-of-change proportional to frequency; this causes the increase in inductive reactance with frequency is referred to as impedance. The formula for inductive reactance is:

$$X_L = 2\pi (3.14) \times (f) \text{ frequency} \times (L) \text{ Inductance in Henrys}$$

The inductive coil has a low measured resistance until it is energized and increases in impedance during operation, due to reactance. Transformers, solenoid coils, and motor windings are examples of components that produce inductive reactance. These devices produce a magnetic field and voltage of their own in direct opposition to the supply voltage. This counter-EMF acts like an additional resistance impeding current flow and is created only when the device is operating. Counter-EMF decreases current flow after start-up and during operation of the device. (Ohms Law: increased resistance decreases amperage, and lower resistance increases amperage.)

CAPACITIVE REACTANCE

A capacitor is a device that stores electrical energy for later use. A capacitor is composed of two conductive plates with an insulating (dielectric) material between them. In an A/C circuit, the capacitor continuously charges and discharges, creating an opposition to current flow or a type of resistance (impedance) referred to as capacitive reactance.

IMPEDANCE

Impedance is the total opposition to alternating current flow. "Z" is the symbol for impedance.

MEASURING RESISTANCE

Ohmmeters are used to check resistance. Ohmmeters are very sensitive and measure resistance in Ohms. They measure electron movement calibrated to a voltage supplied by the meter's battery.

When using an ohmmeter, supply voltage must be turned off and disconnected from the device being tested. When devices are wired in parallel, a conductor should be removed from the device being measured in order to prevent a feedback circuit.

 Failure to disconnect the device from the circuit can cause bodily harm to the technician and/or severe damage to the ohmmeter and can result in false readings caused by electron flow through another circuit.

Resistance readings also reveal specific situations such as continuity, an open circuit, or a short circuit.

Continuity describes a complete path for electron flow and is indicated by zero resistance on an ohmmeter. Continuity indicates no broken wires, open switches or blown fuses.

An open circuit describes an open switch, blown fuse, broken wire, etc. Electrons cannot flow in an open circuit. The ohmmeter reveals an open circuit by unlimited resistance (infinity) or extremely high resistance (megaohms) (O.L. on an ohmmeter).

A short circuit is a complete circuit (continuity) where none should exist. There is very little or no resistance in a short circuit. The ohmmeter detects a short circuit by indicating zero resistance between two points that should indicate a measurable resistance or infinity.

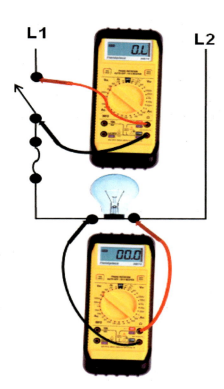

Fig. 1-9: The top ohmmeter indicates an open circuit. The bottom meter shows a shorted circuit.

WATTAGE

James Watt (1763-1819) discovered the method we use for measuring electrical power. Electrical power is the rate at which electricity is used to perform useful work. This work is measured in units called watts. In an actual electrical circuit converting electrical energy to useful work is not 100% efficient. Some of the power produces heat or another byproduct which must be accounted for and is calculated as reduced Power Factor.

Simply put Watts are calculated by multiplying amperage x voltage:

$$P \text{ or } W = I \times E$$

(746 watts = 1 horsepower)

A wattmeter is normally located at the power entry to a building and measures the number of kilowatts (1000 watts = 1 kilowatt) consumed. When calculating the actual energy use the formula Watts = Current x voltage x power factor is used.

Figure 1-10 is a wheel showing the formulas for calculating volts, amperes, resistance, and power. It is a combination of the Ohm's Law circle and the Power Law circle. If any two factors are known, the others can be calculated.

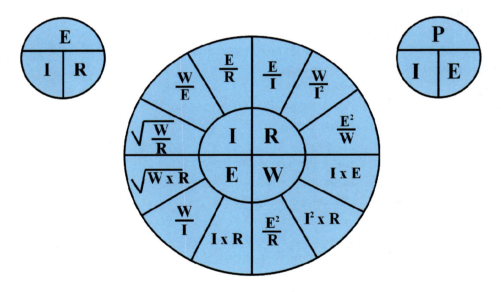

I = intensity (amperes)
R = resistance (ohms)
E = electromotive force (voltage)
W = watts (power)
P = power (watts)

Fig. 1-10

Student Worksheet

Chapter 1: What Is Electricity?

Name _____   Date _____

Solve For Ohm's Law and Power Laws.

1. Find voltage when resistance is 20Ω and current is 6 Amps.

2. Find current when voltage is 240 and resistance is 60Ω.

3. Find resistance when voltage is 24 and current is 3 amps.

4. Find power when current is 10 amps and voltage is 120.

5. Find power when current is 5 amps and resistance is 5Ω.

6. Find voltage when power is 1000 watts and current is 8 amps.

7. Find resistance when power is 960 watts and voltage is 12.

8. Find current when power is 500 and voltage 50.

9. Find power when resistance is 24Ω and current is 10 amps.

10. Find power when voltage is 200 and resistance is 400Ω.

Student Worksheet　　Page 13

Chapter 1: What Is Electricity?

Name _____　　　　Date _____

Use an ohmmeter to measure the resistance of an incandescent light bulb (hot and cold), a heating element, relay coil and contacts (NO and NC), and fan motor. Record measurements.

NOTE: Disconnect and remove all items to be tested from any live circuit. Never use an Ohmmeter with voltage.

1. **Incandescent lamp, cold:** _____

2. **Incandescent lamp, hot:** _____

3. **Heating element:** _____

4. **Relay coil:** _____

5. **Contacts:** _____

6. **Fan Motor:** _____

Electrical Theory & Applications for HVACR　　　　　　　　　　　　　　　　　　　　　©2012 ESCO Group

Student Worksheet

Page 15

Chapter 1: What Is Electricity?

Name _____   Date _____

Use a volt meter and measure the voltage of a 120v receptacle, 240v receptacle, and 24v side of a transformer. Record measurements.

1.   **120V Receptacle**

2.   **240V Receptacle**

3.   **240V Disconnect**

4.   **24V Transformer**

Electrical Theory & Applications for HVACR   ©2012 ESCO Group

Chapter 2: Circuits and Their Components

OBJECTIVES:

- Understand different units of measurement
- Identify series, parallel, and combination circuits and find voltage, current, resistance, and wattage
- Identify the difference between single- and three-phase circuits, and understand their applications
- Recognize electrical symbols
- Understand the procedures for electrical measurements such as voltage, current, resistance, and wattage
- Know the differences between conductors, insulators, and semi-conductors
- Understand wire sizing
- Define protection devices and testing procedures
- Explain the differences between loads and switches
- Know testing procedures for loads and switches
- Explain the differences between step-up and step-down transformers
- Describe the function of a solenoid
- Know the procedures for testing transformers
- Explain the differences between relays contactors and motor line starters
- Know the procedures for testing relays contactors and line starters
- Explain functions and test procedures for defrost timers
- Describe the different thermostats and their functions, as well as test procedures

2 Circuits and Their Components

SIMPLE CIRCUIT

A simple circuit contains a power supply and a load. It has one path of electrons that flows to the load and back to the source. It may or may not contain a switch.

SERIES CIRCUITS

A series circuit has one single path for current flow. Components in the circuit are arranged so that current must flow through the first device, then the second, third, and so on. If the connection is broken or if one device fails, current flow stops in the entire circuit. Voltage is shared (divided) by all devices in a series circuit. As current flows through loads wired in series, the voltage drops as each load consumes power. Switches and contacts are not power-consuming devices and should not cause a voltage drop.

There may be several switches in series with a motor or other device. If any one of the switches opens, current stops.

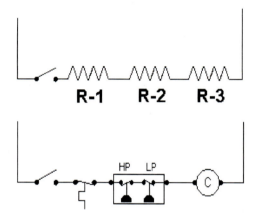

Fig. 2-1: Examples of series circuits

Total Resistance in a Series Circuit

A series circuit has only one path for current flow. Therefore, total resistance is the sum of all of resistances in the circuit.

RT = R1 + R2 + R3...

Series Circuit Laws

1. Only one path for current to flow.
2. Current remains constant throughout the circuit.
3. Total resistance is equal to the sum of all resistors in the circuit.
4. Total voltage is equal to the sum of all voltage drops in the circuit.
5. Largest voltage drop always occurs across the highest resistance in the circuit.

PARALLEL CIRCUITS

A parallel circuit has more than one path for current flow. Parallel circuit configurations are used to connect several loads across the same voltage source. Current flows through each load independent of the others.

Current flow through each load is not necessarily equal, but voltage across the load is always equal.

Remember, voltage (E) remains constant and resistance total is always less than the least resistance in the circuit.

Total Resistance in a Parallel Circuit

Since a parallel circuit has more than one path for current flow, adding additional paths (loads) will decrease the total resistance in the circuit. The total resistance is always less than the smallest resistance in the circuit.

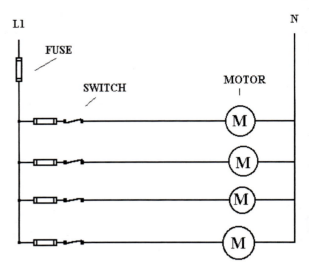

Fig. 2-2: Parallel circuit

Formulas to calculate total resistance in a parallel circuit:

For two resistors:

$$R\ Total = \frac{R1 \times R2}{R1 + R2}$$

For three or more resistors:

$$R\ Total = \frac{1}{\frac{1}{R1} + \frac{1}{R2} + \frac{1}{R3}}$$

Parallel Circuit Laws
1. More than one path for current to flow.
2. Voltage remains constant throughout the circuit.
3. Total current is equal to sum of all branch currents.
4. Total resistance is always less than the least resistor in the circuit.

COMBINATION SERIES/PARALLEL CIRCUITS

Voltage, resistance, amperage, and power in a complex circuit such as a combination series/parallel circuit vary in each section of the circuit. To determine these values, the parallel and series sections of the circuit must be calculated separately before finding circuit total. In Figure 2-3, the parallel section as shown in the circle must be calculated before adding it to the series section of the circuit.

THREE-PHASE CIRCUITS

A power plant generator rotates a magnetic field within three different conducting loops at the same time. This is called poly-phase generation, and it is much like having three different power plants. The three conducting loops are spaced exactly 120 degrees apart and are called phases (Ø) or legs. While one phase is magnetically positive, another phase is negative and the other is at zero. These three phases take turns changing polarity from positive to negative to zero at a rate of 7,200 times per minute. This three-way positioning is continuous as each loop rotates inside the generator, producing alternating current in each of the three phases that are out-of-step or out of phase with each other. The voltage variance between phases should not exceed two percent.

COMPLEX CIRCUIT

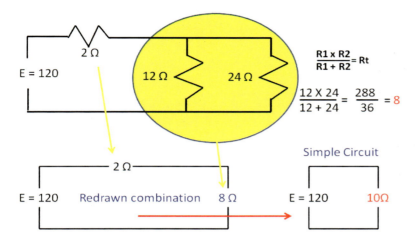

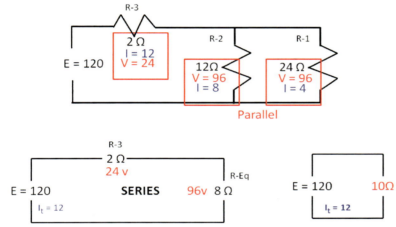

Fig. 2-3: Complex circuit broken down to its simplest form

A potential difference exists between any two hot wires because their polarity is different. Each hot wire has the same voltage, but different polarity. Electrons flow according to polarity, from negative to positive. Therefore, the potential between any two hot wires is additive, because 120 volts (negative) plus 120 volts (positive) equals 240 volts. One wire is "pushing" while another is "pulling" with equal force. This is called alternating current (AC) because the hot wires are alternating between positive and negative at a rate of 7,200 times per minute, or 60 cycles. The electrons are simply changing the polarity. A voltage tester reads the effective voltage of the two wires; this is called the root mean square (rms). See Figure 2-4a.

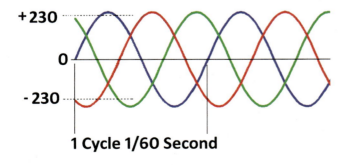

Fig. 2-4a: 3-phase sign wave, each leg 120° apart

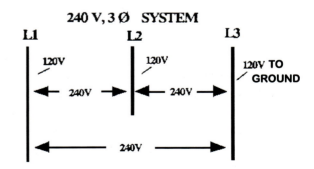

Fig. 2-4b: Voltage between any two wires is additive because polarity is different

Many electrical loads, generally commercial, are designed for connection to all three hot wires; these are called three-phase loads. The normal color code for these wires is black or red, but they can be any color except white or green. The letter "L" for "line" is used to help identify the three phases (L1, L2, L3).

SINGLE-PHASE CIRCUITS

Some loads are designed to operate with just two hot wires from a three-phase system. These are called single-phase loads, not two-phase. (The term "two-phase" refers to an old system still used in a few remote areas.) Higher voltage is obtained by using two wires, from a three-phase system, that alternate from positive to negative. This push-pull effect can be obtained with any two phases. See Figure 2-5. The allowable voltage variance in a single-phase circuit is ten percent.

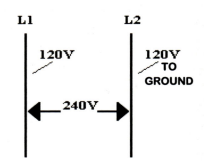

Fig. 2-5: 240-Volt, 1Ø System

THE NEUTRAL WIRE

Earth is an excellent conductor of electricity, and damp soil is a better conductor than dry soil. The earth is always at zero potential (no voltage) and can be used to complete an electrical circuit. Many electrical devices operate with one hot wire from a three-phase system and another wire called the neutral. This method is also called single-phase, and involves lower voltage. A potential difference exists because the hot wire has voltage and polarity, but the neutral wire is connected to the earth, or grounded, which has zero voltage.

While the hot conductor usually has black insulation, it can be another color (except white or green) for ease of identification. The neutral wire has white or gray insulation and is connected to a solid copper rod driven eight feet into the ground. This copper rod is called a "grounding electrode." This grounded neutral wire provides a pathway for electrons traveling to and from the Earth, and has zero voltage. The neutral wire is a current carrying conductor, but has no voltage to ground. A potential difference exists between the neutral wire (no voltage) and the black wire having 120 volts. See Figure 2-6.

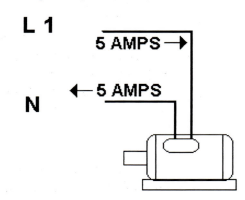

Fig. 2-6: Amperage flowing in the hot wire also flows in the neutral wire

THE SAFETY GROUND WIRE

The equipment grounding conductor is another wire added for safety purposes. This is called the safety ground and is required by the National Electric Code on all new electrical systems. The color code for this wire is green or bare copper (not insulated). The safety ground, or chassis ground, connects to the same terminal as the neutral wire once at the service panel. The neutral wire normally carries current. The safety ground only carries current in the event of a short circuit. This conductor serves strictly as a safety valve when a ground fault occurs in the electrical equipment. The safety ground wire is connected to the frame of a motor or appliance and provides an alternate pathway for electrons to travel to ground and not through someone's body. Many commercial and industrial applications

Fig. 2-7: Motor safety ground

require all electrical wires to be installed inside metal conduit pipes (not plastic). This metal-to-metal pathway should be grounded by connection to a grounding electrode or to the steel framework of the building (which is grounded).

CONDUCTORS

Electrical wires are used as conductors to provide the necessary free electrons and serve as a pathway for electron flow. These wires are used to connect devices and switches to complete an electrical circuit. Silver, copper and aluminum are good conductors because they have high conductivity and low resistance to current flow. In general, any material that has three or less electrons in its outer orbit, called the valance ring, is considered a good conductor. Copper is the most commonly used conductor. Copper has a single electron in its valance ring.

The material's ability to conduct electricity is referred to as its K Factor. (Conductivity.)

The amount of current a conductor can safely carry without becoming overheated is limited. This current-carrying ability is called ampacity. The ampacity of a conductor depends upon the wire's diameter, length, location, type and quality of insulation. The chart shown in Figure 2-8 is a partial list of wire sizes, resistance, and ampacity for standard sizes of copper and aluminum wires based on American Wire Gage (AWG).

Consult the National Electrical Code (NEC) book for a complete list of up-to-date information. No. 12 copper wire is probably the most commonly used wire size. It is often used when a smaller wire would be approved. Other than cost, there is no problem with over-sizing a wire. Under-sizing causes severe problems due to overheating. Figure 2-9 illustrates a sample of various wire sizes for both solid and stranded wires.

Gauge No. (AWG)	OHMS per 1000 FEET	Ampacity Copper	Ampacity Aluminum
0000	0.050	230	180
000	0.062	200	155
00	0.080	175	135
0	0.100	150	120
1	0.127	130	100
2	0.159	115	90
3	0.202	100	75
4	0.254	85	65
6	0.40	65	50
8	0.645	50	40
10	1.02	30*	25
12	1.62	20*	18
14	2.57	15*	
16	4.10	10*	
18	6.51	5*	

*Load current rating and over-current protection shall NOT exceed these figures.

Fig. 2-8: Ampacity of commercial wire

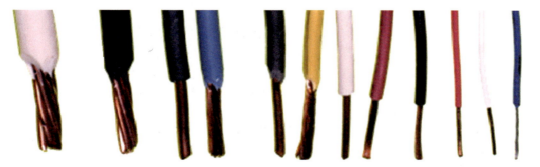

Fig. 2-9: Wire size and type determines current carrying ability (ampacity)

INSULATORS

Insulators are materials that offer high resistance to electron flow. Materials that have 5 or more electrons in the outer orbit are considered insulators. There is no perfect insulator. Insulators can break down due to moisture, heat, excess current flow, vibration, chemicals, etc. Insulation can be heat resistant, moisture resistant, oil resistant, etc. The type of insulation or covering determines where the conductor can be safely used. Always use care to avoid damaging the insulated covering on the wire.

The manufacturers of electrical wires use letter codes to designate the type of insulation on a wire. Insulation types are coded by letters of the alphabet and stamped on the insulation surface. The most commonly used types are TW, THN, TFF, or THWN.

Consult the NEC handbook for a complete list of insulation types. Figure 2-10 shows how conductor insulation is marked.

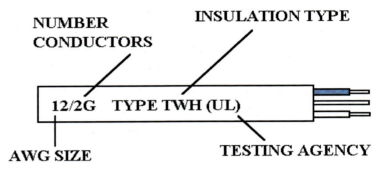

Fig. 2-10: Insulation information

SEMI-CONDUCTORS

A semi-conductor is a material that has electrical properties of current flow between a conductor and an insulator (four electrons in the valance ring.) Silicon is an example of such material. Pure silicon is not a good conductor because it has four electrons in its outer orbit that bond with other silicon atoms to form a stable crystal.

The outer ring of a silicon atom has four electrons, but there is room for eight. Electrons share orbits with other atoms to form covalent bonds. If an impurity with either three or five electrons in the outer orbit is added to silicon, the crystalline structure will have either an excess electron or a hole and can become a conductor or insulator if the correct voltage and polarity are applied. The new structure is called P-Type or N-Type material.

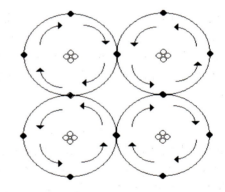

Fig. 2-11: Silicon atoms

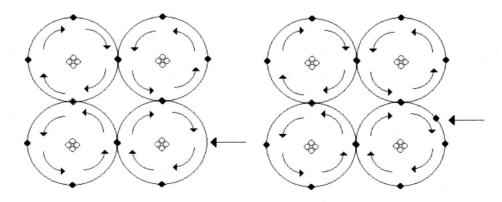

P-Type Material (Missing 1 Electron)   N-Type Material (1 Extra Electron)

Fig. 2-12: P-Type and N-Type material

An electrical check valve can be created by sandwiching pieces of N-Type and P-Type material together. Electrons can flow into the N-type material and out of the P-Type material, but electrons attempting to enter the P-Type material are blocked and no current flows. This simple solid state device is called a diode. A diode is represented in electrical schematics by the symbol in Figure 2-13.

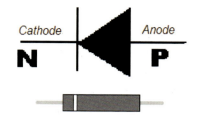

Fig. 2-13: Schematic symbol for diode

CIRCUIT PROTECTION

Overcurrent occurs when excess current is flowing through a wire. This causes the wire to become hot and presents a serious hazard. Overcurrent can be caused by a variety of electrical problems such as loose connections, ground fault, short circuit, defective resistance, or too many loads. An overload describes an overcurrent (between two and ten times the rated current.) A ground fault describes an overcurrent which runs to ground and may be hundreds of times the rated current. A ground fault is VERY dangerous.

Fuses and circuit breakers protect a circuit against overcurrent. The amperage rating of a fuse must not be greater than the ampacity of the wires being protected. Fuses and circuit breakers are used to detect excessive load current and open the circuit before danger arises. Examine the fuse clamp holder for discoloration or loose connections; this indicates an overheated connection. It is standard practice to locate fuses in the main power supply and in each branch circuit. A blown fuse in a branch circuit helps confine the problem to a specific area. Fuses, overloads, and circuit breakers protect wires and equipment, not people.

The amperage and voltage rating of a cartridge fuse determines the physical size of the fuse. Fuse holding devices are sized according to the same procedure. This helps prevent oversizing fuses on circuits designed for a certain maximum amperage. Figure 2-14 illustrates fuse dimensions in inches according to voltage, and amperage ratings of the fuses.

AMPERAGE RANGE	250 V FUSE LENGTH (INCHES)	600 V FUSE LENGTH (INCHES)
1/10 to 30	2	5
35 to 60	3	5 1/2
70 to 100	5 7/8	7 7/8
110 to 200	7 1/8	9 5/8
225 to 400	8 5/8	11 5/8
450 to 600	10 3/8	13 3/8

Fig. 2-14: Amperage and voltage determine fuse size

Cartridge fuses are available as ordinary fuses (one-time blow) or dual-element. Dual-element, or time-delay, fuses permit an overload of short duration, but blow instantly if a short circuit occurs. Time-delay fuses are necessary when fusing circuits for electric motors. Screw-in or time-delay fuses are sized up to 30 amperes. If the view port on a screw-in fuse is blackened or the skinny part of the element is burned, the fuse has experienced extreme overload. If a port reveals a collapsed spring inside, the fuse has experienced a slight overload. Round cartridge fuses are used up to 60 amperes and knife-blade contacts are used for fuses over 60 amperes. It is important that fuse ends make tight contact in the fuse holder. Loose connections or high air temperatures around a fuse reduce the amperage rating and cause needless shutdowns. Remember never to pull a fuse under a load, and always use a fuse puller to remove cartridge fuses.

LOADS AND SWITCHES

Manufacturers of electrical devices install the correct type of resistance for the device to perform the correct amount of energy conversion. It is important to connect these loads to the designed voltage. When connecting a load to a voltage source, a minimum of two conductors must be used. A potential difference (voltage) must exist between the two wires. This power source is connected to each end of the resistance. The potential difference causes electrons to flow through the resistance. Electrons flowing through the resistance causes electrical energy to be converted to another form of energy to provide useful work.

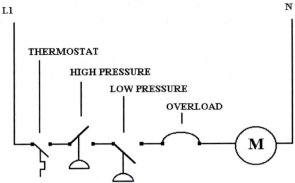

Fig. 2-15: Multiple switches safeties

A load cannot operate unless the circuit provides a complete pathway for electrons to flow into and out of the load. Switches are always connected in series with a load (one after the other). More than one switch is often used to control and/or provide safety protection.

To provide proper voltage supply, energy conversion devices, are usually connected in parallel. A parallel circuit is connected "from one side of the power supply to the other". The ultimate test of a parallel connection is that the device can be removed without effecting the operation of other devices. In parallel connections, each device is connected independently from all others.

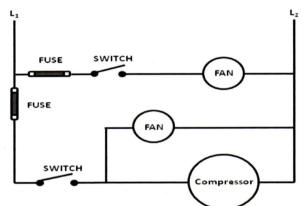

Fig. 2-16: Switches and fuses connected in series with loads

When more than one load is connected to a power source, switches are required to control the individual loads. The switches are connected in series with the loads and each load is connected in parallel to the power supply. Sometimes one switch may control more than one load. See Figure 2-16.

TRANSFORMERS

A chief advantage of alternating current is that it can be generated at one voltage, transmitted at a higher voltage, and then reduced to a lower voltage at the point of use. Transformers make it possible to increase (step up) or decrease (step down) the voltage. When voltage is stepped up, the required wire size for the secondary side is reduced, decreasing transmission cost. A transformer has two windings: a primary (incoming voltage) and a secondary (outgoing voltage) winding. Voltage at the secondary is determined by number of coils or wraps in the secondary winding versus number of coils in the primary winding. Transformer primary terminals are normally labeled H1, H2, and H3, and secondary terminals are tagged X1, X2, and X3. A neutral terminal is labeled X0.

Single-phase transformers are rated by VA (volts x amps) at the secondary winding. Transformers rated over 1,000 VA are usually rated KVA, with K representing 1,000. An overloaded or undersized transformer will burn out because the secondary coil cannot carry the required current. The secondary winding is considered to be a power source for any loads connected to the transformer. Single-phase transformers usually have two wires for the primary winding and two for the secondary.

Three-phase transformers can be of the Wye or Delta type, although the Wye, or Star, is most popular. Because electricity is produced in three phases, a variety of Wye, Delta, Wye-Delta, or Delta-Wye transformers are used. Various voltages are in common use. See Figures 2-17(a-d).

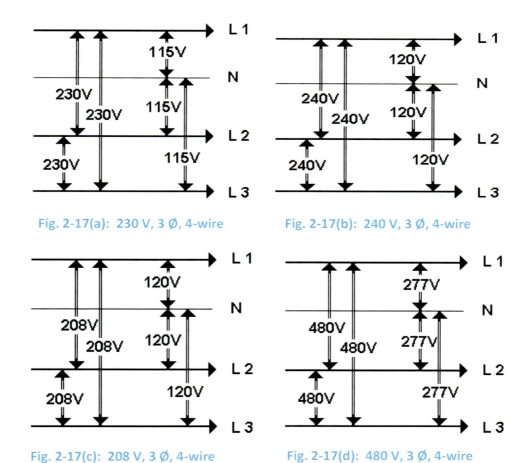

Fig. 2-17(a): 230 V, 3 Ø, 4-wire
Fig. 2-17(b): 240 V, 3 Ø, 4-wire
Fig. 2-17(c): 208 V, 3 Ø, 4-wire
Fig. 2-17(d): 480 V, 3 Ø, 4-wire

Some older systems use a high-leg system from a three-phase (3Ø) Delta transformer. Voltage readings from two of the hot legs to neutral will read 115 volts. However, one of the hot legs to neutral will register 208 volts. This wire is called the stinger, high leg, or wild leg, and cannot be used for 115-volt circuits. See Figure 2-18.

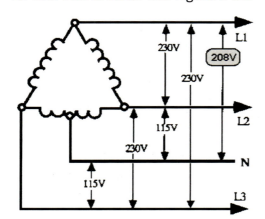

Fig. 2-18: Delta high-leg system

Fig. 2-19(a): Single-phase 230/120 V

Fig. 2-19(b): 230/120 V three-phase with high-leg

Fig. 2-19(c): 230/120 V three-phase

LOW-VOLTAGE TRANSFORMERS

Step-down transformers have less windings in the secondary than the primary, which makes resistance less in the secondary. A 240-volt primary has a ratio of 10:1 windings for a 24-volt secondary; a 120- to 24-volt has a ratio of 5:1 for 24 volts.

Twenty-four-volt control transformers used in HVAC equipment have a volt-amp rating between 16 and 50. To find the amperage capacity of a low-voltage transformer, divide the volt/amp (VA) rating by the secondary voltage output.

Example:
Rating: 24 volts/40 VA
40 VA ÷ 24 volts = **1.7 amps maximum secondary load**

Most transformers have an internal fuse to prevent the transformer from high-amperage burn-out. Some manufacturers install a ±2 amp time-delayed fuse in the secondary side for additional protection.

Some equipment has two transformers: one in the indoor unit and one in the outdoor system. Depending on equipment and type of controls used, transformers may need to be wired in phase to each other or in phase to ground.

To properly wire two low-voltage transformers in phase, the primary and secondary windings of both transformers must be connected so that current flow is in the same direction in each transformer. L1 and L2 must be connected as shown in Figure 2-20(a) on the transformer, L1 to L1 and L2 to L2.

The secondary windings of both transformers must be connected in parallel and in phase. To check the phasing connect one 24-volt lead from each transformer together. With a voltmeter, measure the voltage between the two other 24-volt connections. If in phase, the voltmeter will indicate 0 volts; if out of phase, the voltmeter will indicate 48 volts. If the measurement is 48 volts, the wires on either the primary or secondary side of the transformer must be reversed.

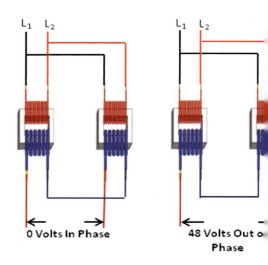
Fig. 2-20(a): Transformer phasing

When it is required that one of the leads of a low-voltage circuit is grounded to the equipment, the grounded lead must be in phase with the ground. (This configuration is common on equipment with electronic controls.)

Voltage between the line voltage feed (L1) and the 24-volt lead used for the feed should be lower than the supply voltage, as shown in Figure 2-20(b).

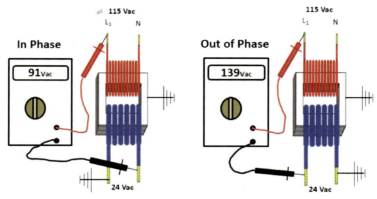

Fig. 2-20(b): Transformers connected

The standard wire size for field assembled low-volt control electrical wiring is 18-gauge for runs of 50 feet or less. Follow all manufacturer recommendations.

SOLENOID VALVES

Electrically operated valves are called solenoid valves. When current flows through the coil, electromagnetism is produced and lifts the valve's plunger. This opens the valve and liquid flows through the valve. When the coil is de-energized, the magnetic field collapses (disappears) and the valve closes (normally closed). Some solenoid valves are designed to be normally open, and will close when the coil is energized. Solenoid valves are frequently used in air conditioning and refrigeration for hot gas bypass systems and automatic pump-down.

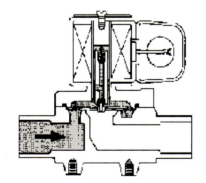

Fig. 2-21: Solenoid valve

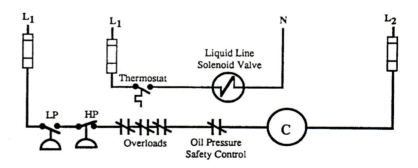

Fig. 2-22: Pump-down circuit

RELAYS

A relay uses a magnet to operate a switch or contacts. Relays are often used to control electrical loads from remote locations. When a relay coil is energized, electromagnetism causes a switch to open or close. The electrical circuit to the relay coil is entirely separate from the circuit through the relay contact. The coil voltage may be 24 volts and the circuit through the relay contacts may be 120 volts. Thus, low voltage is used to control a switch that controls a high-voltage load. See Figure 2-23.

Relays often contain more than one set of contacts or switches. Relay contacts are always shown as normally open (NO) or normally closed (NC) with the coil in the de-energized position. Energizing the coil causes the contacts to change position. The relay contacts have low current ratings, with a maximum of 10 amps being considered normal. Current flow in the coil circuit is very low, often less than 1/4 ampere. A 24-volt control circuit is safer for personnel, permits the use of smaller wire, and causes less arcing at the switch.

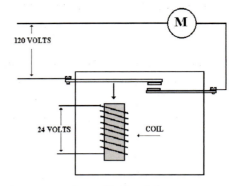

Fig. 2-23: The coil circuit is separate from the motor circuit

A lock-out relay is a relay with approximately 75 percent more resistance than the coil of the control it locks out. This difference allows for the proper voltage drop so that one coil will energize and the other will not. It is wired in a circuit as shown in Figure 2-24.

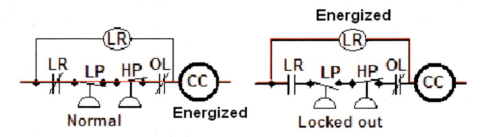

Fig. 2-24: Lock-out relay circuit

SEQUENCER

If multiple loads are energized at the same time, an overwhelming current surge or load is placed on the power supply. When this occurs in a residential home, lights may dim or flicker; in a commercial building, electrical demand for purchased electrical energy will be billed at a high-demand rate for the entire month.

Sequencers are time-delay relays that stage loads on and off and are commonly found on electric heating systems connected to heating elements.

CONTACTORS

A contactor is an electrical switching device that operates much like a relay. However, the contactor has heavy-duty contacts for controlling larger loads. These contacts are normally open (NO) and controlled by a magnetic coil. When the coil is energized, the contacts close. Contacts are used to open or close a circuit between the main power supply and the load and are rated according to the maximum amperage flow through the contacts for a specific voltage. A two-pole contactor has two separate contacts and is used to control 240-volt single-phase circuits (for residential air conditioning). A three-pole contactor has three sets of contacts and is used to control three-phase loads (for commercial and industrial).

It is normal for contacts to become pitted and burned due to arcing when they open and close. Use of a file or sandpaper to clean the contacts is not recommended; such "cleaning" destroys contact surfaces and increases arcing. Large amperage replacement contacts are usually available from local suppliers.

The numbering system for contactors determines the direction of current flow through the contacts. It is standard policy for power supply to enter at the top of the contacts and the load to be connected to the bottom. Line power (inlet) terminals are labeled L1, L2, and L3 and load (outlet) terminals are labeled T1, T2, and T3, according to the number of poles.

Fig. 2-25: Single-pole contactor

CONTACTOR COIL BURN-OUT

The magnetic coil is located inside the contactor but is not electrically connected to the main contacts. Because the coil is a separate device, coil voltage can be different than voltage at the main contacts. The coil has its own terminals for making electrical connections, and replacement of the coil is easy. Coils may be 24, 120, or 230 vac. And contacts may be rated 15 to 60 amperes. Because the coil is electrically separate from the main contacts, it is common practice to use lower voltage to operate the coil. Residential systems generally use 24 vac. control system. Current in the control circuit is safer and permits the use of standard switching devices and class-two thermostat wire. A step-down transformer is used to obtain lower control circuit voltage.

Any number of contacts and safety control switches can be located in the control circuit. Contacts for these controls are connected in series with the coil and can include overloads, thermostats, pressure controls, fuses, limit switches, flow controls, or oil pressure controls. All switches in the control circuit in Figure 2-26 must be closed before the coil can be energized. However, any switch can disconnect power to the coil and stop the motor. Some safety controls have a manual reset, while others have an automatic reset.

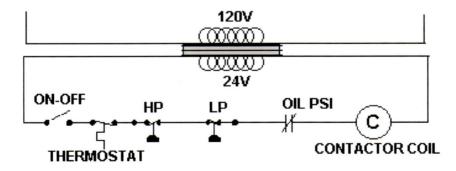

Fig. 2-26: All control switches must be closed to energize the coil; any switch can open and stop the motor

LINE STARTERS

A line starter is basically a contactor with built-in overload protectors. Line starters are often used to operate and protect three-phase motors. Overloads protect the motor against excess amperage and are more accurate than fuses. Overload protectors are normally connected to the bottom of the contactor and sized to the amperage of the motor. One overload is required for single-phase power supply to the motor. For a three-phase line starter, there must be at least two overloads. Supply voltage must flow through heavy-duty contacts and then through the overload heater before traveling to the motor.

Overload heaters are connected in the high-voltage power supply in series with the motor. They are sized to permit a specific amount of amperage before producing heat. Excess amperage causes the heater to generate heat that causes a nearby set of control contacts to open. Overload contacts are connected in series in the control circuit that supplies power to the contactor coil. Figure 2-27 shows a cut-away view of one leg of the contactor circuit and the connecting overload.

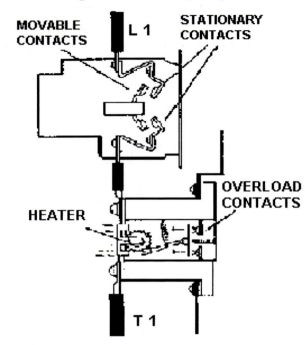

Fig. 2-27: Cutaway view of contactor and overload

A manual reset is provided to re-close overload contacts after the heater cools off. Heaters are available in a wide variety of amperage capacities and are normally sized to provide accurate motor protection at slightly above full-load amps (FLA).

Three-phase power supply to the motor is connected through heavy duty contacts, the heater, and then the motor. Overload switches are connected in the control circuit to the contactor coil. Excess amperage in a high-voltage leg causes the overload contacts to open. Opening overload contacts disconnects the power supply to the contactor coil, stopping the motor. A tripped overload switch usually requires manual reset of the contact; any of the three heaters can open the contact. The line starter also has one or two sets of auxiliary contacts, numbered 1, 2, and 3; one set is closed and one open. These contacts can be used for many purposes, like connection for start/stop switches, interlock switches, and two power indicator lamps. It is not uncommon for a line voltage starter to have a switch on the cover for selecting manual and automatic operation.

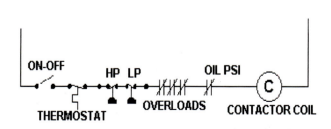

Figure 2-28: Overload circuit protection

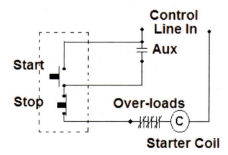

Figure 2-29: Line voltage starter with start/stop switch

DEFROST TIMERS

To provide automatic defrosting of the evaporator coil, an electric heating element energized by a defrost timer may be located near the evaporator. A defrost timer is operated by a synchronous motor like those used to operate wall clocks. A cam that is gear-driven by the motor operates a set of electrical contacts. At the proper time, these contacts change position, stopping the cooling process and energizing the defrost heater or hot gas solenoid valve.

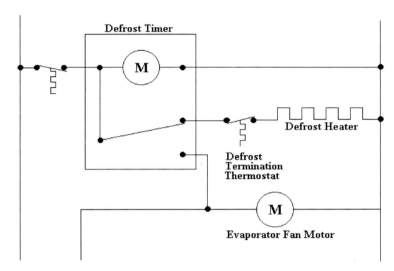

Fig. 2-30: Domestic refrigeration defrost timer circuit

The synchronous motor in the timer can be wired in two ways: as a continuous-run timer or as a cumulative compressor-run timer. In the continuous-run mode, the timer is wired directly across the power source and operates on a continuous basis. In cumulative compressor-run mode, the timer is wired to operate only when the compressor is in operation and the thermostat is closed. Although defrost function is generally an issue in refrigeration units, heat pump air conditioning systems require periodic defrosting of the outdoor coil under certain conditions. Electronic defrost controls may be used in place of mechanical controls, providing precise control without moving parts. Defrost timer controls for commercial equipment have a means to automatically advance the timer back to the run cycle before the time sequence if defrost temperature is satisfied.

THERMOSTATS

A thermostat is a temperature-sensitive switch. Some thermostats are designed to operate at low voltage, generally 24 volts, while others are designed for high voltage and are connected directly to motors or heating units. Low-voltage thermostats are more economical and safer to use inside the home.

A residential air conditioning system thermostat controls three major components:
- Compressor
- Condenser fan (comes on with the compressor)
- Evaporator fan motor or blower

A residential heating system thermostat controls one major component:
- Heating (sequencer, gas valve, or electronic control board)

A heating thermostat opens on temperature rise and closes on temperature drop. Conversely, air conditioning thermostats close on temperature rise and open on temperature drop.

One common type of temperature sensing thermostat uses a bimetal switch. Two dissimilar types of metal are bonded together to form a bimetal. Because the metals expand at different rates, a change in temperature will cause the bimetal strip to bend. A contact switch at the end of the strip can then be either opened or closed, depending on the application. Most often, the bimetal strip is bent into a helix to save space.

Another type of switch that has been used with bimetal is a mercury contact. A small pool of mercury is sealed inside a glass container that also contains a set of contacts. Most mercury switches are constructed as single-pole double-throw, which means there is a common terminal, a normally closed terminal, and a normally open terminal. This design allows the thermostat to be used for both heating and cooling applications. Mercury has been phased out due to its toxicity.

Newer electronic programmable thermostats have many more features. An advantage of programmable thermostats is the ability to raise or lower a structure's temperature at pre-specified times on each day of the week.

Electronic thermostats are very versatile, providing multiple options for heating and cooling control. The same thermostat can be programmed to operate a fossil fuel system as well as a heat pump system with multiple stages of heating and cooling. Electronic thermostats do not use an heat anticipator. Most have an option to set cycle time per hour. The thermostat will adjust run time based on a biorhythm developed by the micro processor.

Some heating and cooling thermostats using a single transformer have a jumper connection on the sub-base of the thermostat to tie the circuits together. The thermostat may have the letters RH or the number 4 to indicate the feed for the heating circuit and R or RC for the cooling circuit feed.

For proper operation a residential heating/air conditioning system thermostat should be located at the height that local codes or ADA regulations mandate, in a place where drafts or sunlight will not affect its temperature sensor.

Thermostats must be mounted level to operate effectively. Failure to mount a thermostat correctly will result in incorrect temperatures in the conditioned space. The small hole in the wall behind the thermostat for the entrance of thermostat low-voltage wiring MUST be sealed or poor temperature control will result and room temperature will not be maintained to the thermostat set point.

HEAT ANTICIPATORS

Mechanical bimetal thermostats have a fixed anticipator for cooling and an adjustable anticipator for heating. The heating anticipator is an adjustable resistance heater located near the bimetal coil in a thermostat. It heats the bimetal strip slightly to prevent overshoot of the system. For example, a forced-air system operates by heating a metal heat exchanger. When the heat exchanger reaches the operating temperature, a thermostatic or electronic switch closes, turning on the blower motor to move heated air to the living area. When the temperature set point is reached, the thermostat breaks the circuit to the heat source. The blower operates for a longer period of time to cool the furnace. This extra heat from the furnace heats the conditioned area to a higher temperature than the set point of the thermostat. The heating anticipator provides a little heat to the thermostat bimetal, making it break the heating circuit before reaching the room set point and thus preventing overshoot.

A heat anticipator is wired in series with a heating control circuit and is set according to the amperage draw of the control circuit. Technicians must determine the amperage draw (by measuring the amps between the R and W terminals on the thermostat sub-base) while the furnace is operating at a steady state condition. This usually occurs after about one minute of operation. If the anticipator is set at higher amperage, system overshoot will occur; if it is set at lower amperage, system lag will occur or possible anticipator burnout.

The cooling anticipator is a fixed resistor that is in the circuit when the cooling thermostat circuit is open. It is parallel to the cooling contacts. The cooling anticipator provides heat to make the system operate, before the conditioned temperate gets too high and prevents a lag in cooling.

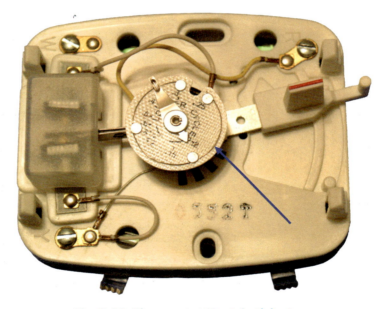

Fig. 2-30: Thermostat Heat Anticipator

Student Worksheet

Page 37

Chapter 2: Circuits and Their Components

Name _____   Date _____

A 10 Ω, 20 Ω, and 30 Ω resistor are connected in series to a 120-volt power source. Find the total resistance, current, voltage drop across each resistor, as well as the power for the following circuit.

Record your answers in the boxes below.

R-1   R-2   R-3

	Resistance (Ω)	Current (I)	Voltage (E)	Power (W)
OHM'S LAW SERIES ORGANIZER				
R-1	10			
R-2	20			
R-3	30			
Total			120	

Electrical Theory & Applications for HVACR

Student Worksheet

Chapter 2: Circuits and Their Components

Name _____   Date _____

Find the total resistance, current, voltage drop across each resistor, and power for the following circuit.

Record your answers in the boxes below.

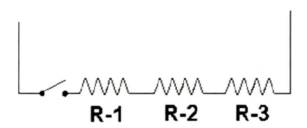

	OHM'S LAW SERIES ORGANIZER			
	Resistance (Ω)	Current (I)	Voltage (E)	Power (W)
R-1	24			
R-2	6			
R-3	10			
Total			240	

Student Worksheet

Chapter 2: Circuits and Their Components

Name _____   Date _____

Find the total resistance, current, voltage drop across each resistor, and power for the following circuit.

Record your answers in the boxes below.

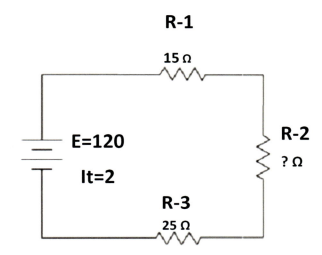

OHM'S LAW SERIES ORGANIZER				
	Resistance (Ω)	Current (I)	Voltage (E)	Power (W)
R-1	15			
R-2	?			
R-3	25			
Total		2	120	

Student Worksheet

Page 43

Chapter 2: Circuits and Their Components

Name _____ Date _____

Find the total resistance, current, voltage drop across each resistor, and power, for the following circuit.

Record your answers in the table below.

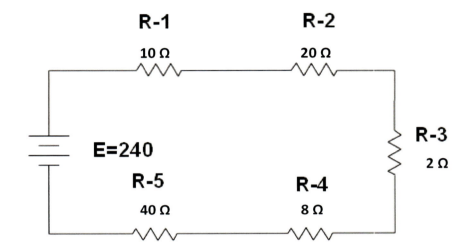

OHM'S LAW SERIES ORGANIZER				
	Resistance (Ω)	Current (I)	Voltage (E)	Power (W)
R-1				
R-2				
R-3				
R-4				
R-5				
Total				

Electrical Theory & Applications for HVACR

Student Worksheet

Page 45

Chapter 2: Circuits and Their Components

Name _____   Date _____

Solve for Power, Voltage Drop, Branch Current, and Total Resistance in the following circuit using the Parallel Organizer. Record your answers in the table below.

Remember, Voltage (E) Remains Constant

Resistance Total is always less than the least (R)

Left to Right Multiply (R x I = E) or (I x E = W)

Up to Down Add ($I_1 + I_2 = It$) or ($P_1 + P_2 = Pt$)

Right to Left Divide (W / E = I) or (E / I = R)

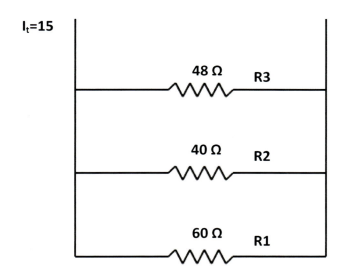

OHM'S LAW PARALLEL ORGANIZER					
	Resistance (Ω)	Current (I)	Voltage (E)	Power (W)
R-1	60			
R-2	40			
R-3	48			
Total		15		

Electrical Theory & Applications for HVACR   ©2012 ESCO Group

Student Worksheet

Chapter 2: Circuits and Their Components

Name _____   Date _____

Find the unknown resistance and solve for power, voltage drop, amperage, and total resistance in the following circuit using the parallel organizer. Total amperage for the circuit listed below is 90.

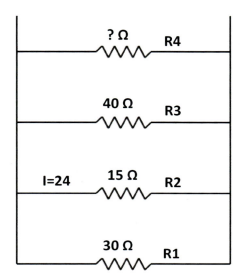

OHM'S LAW PARALLEL ORGANIZER

	Resistance (Ω)	Current (I)	Voltage (E)	Power (W)
R-1	30			
R-2	15	24		
R-3	40			
R-4	?			
Total		90		

Student Worksheet

Page 49

Chapter 2: Circuits and Their Components

Name _____   Date _____

Find total resistance, total current, and total power and solve for current and voltage drop for each resistor. Record your answers in the spaces below.

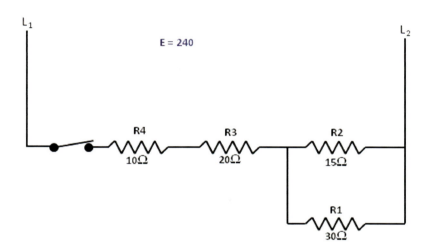

R1 E = _____   **R2** E = _____

  I = _____   I = _____

  R = _____   R = _____

  P = _____   P = _____

R3 E = _____   **R4** E = _____

  I = _____   I = _____

  R = _____   R = _____

  P = _____   P = _____

Electrical Theory & Applications for HVACR   ©2012 ESCO Group

Student Worksheet

Chapter 2: Circuits and Their Components

Name _____   Date _____

Find total resistance, total current, and total power and solve for current and voltage drop for each resistor. Record your answers in the spaces below.

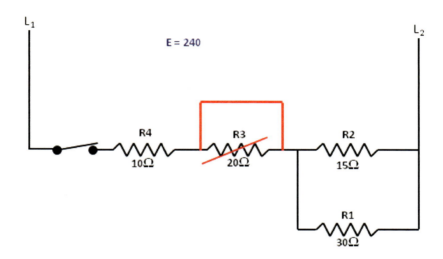

R1   E = _____   R2   E = _____

     I = _____   I = _____

     R = _____   R = _____

     P = _____   P = _____

R3   E = _____   R4   E = _____

     I = _____   I = _____

     R = _____   R = _____

     P = _____   P = _____

Chapter 3: Motors

OBJECTIVES:

- Identify motor components and their functions
- Describe the different types of single-phase motors
- Explain the different motor applications
- Explain motor starting torque
- Calculate motor speed and slippage
- Test motor windings
- Identify run, start, and common windings
- Describe components, functions, and operation of ECMs
- Describe how to connect different motor speeds
- Check run, start, and dual capacitors for shorts, opens, grounds, and mfd ratings
- Be familiar with motors used in modern HVACR equipment, as well as motors that have been used by various manufacturers in the last two decades

3 Motors

Motors are components of circuits, however, this book treats them as a separate section.

INDUCTION MOTORS

Over 90 percent of all motors are induction motors and operate on alternating current. (DC motors are not discussed in this book.) Knowledge of induction motors can easily be applied to other motor types. A constantly malfunctioning motor is a warning that there may be other system problems like improper voltage, faulty capacitor, dirt, moisture, or overloaded conditions. A good working knowledge of motors and their operation is necessary to properly troubleshoot HVACR systems and perform required repairs.

THE PARTS OF A MOTOR

There are three main parts of any motor: the rotor, the stator, and the endbells/bearings. The rotor is located inside the stator. Endbells with bearings are used on each end of the motor and the entire assembly is bolted together.

Induction motors operate on the principle of induced magnetism. The stator is a circular stationary electromagnet, and the rotor is located inside it. When current flows through the stator coils, a strong magnetic field is produced in the stator poles. This stator magnetism induces opposite magnetism in the rotor, causing it to rotate.

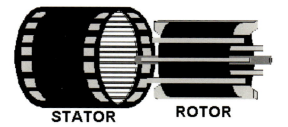

Fig. 3-1: The two main parts of a motor

STATOR POLES

Two or more stationary electromagnets called poles are positioned at opposite sides of a circle inside the motor. A strong magnetic field is produced when current flows through the coils. In a two-pole motor, the stator poles have opposite polarity; one coil produces a north pole and the other a south pole.

An electromagnet has two distinct advantages: its core is magnetized only when current flows through the coil, and its polarity can be changed by reversing this current flow.

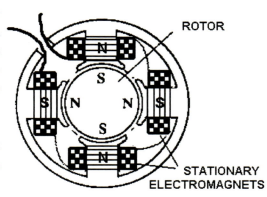

Fig. 3-2: Rotor magnetism is attracted and repelled by rotating stator pole

Alternating current automatically reverses polarity of the stator poles at a rate of 120 times per second (one positive and one negative per cycle). When alternating current reverses, the polarity of each stator changes. Thus, the polarity of stator electromagnets automatically changes with alternating current flow.

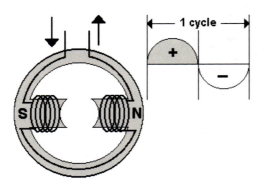

Fig. 3-3: Coils wrapped on stator poles

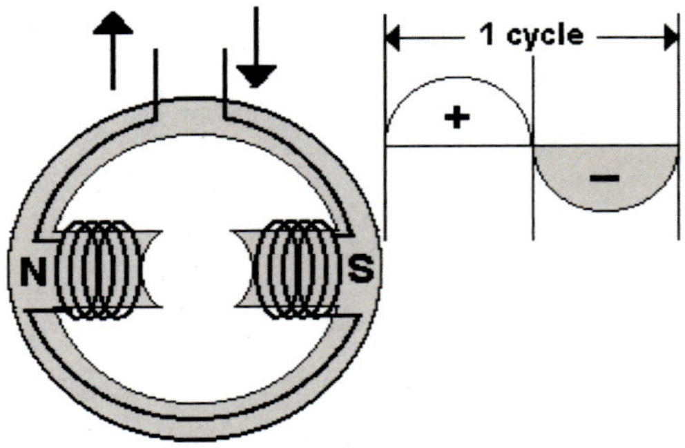

Fig. 3-4: Polarity changes 120 times per second

STATORS

Motors have stationary coils of copper wire called main windings, which are carefully wrapped around layers of soft iron called poles. These magnetic poles consist of coils and laminated cores and are permanently mounted inside of the motor shell. A minimum of two poles, one north and one south, is required. Each pole is located exactly 180 degrees around the circle. This arrangement of magnetic poles is called the stator.

The size (AWG) of copper wire used and the number of wraps in the coil determines the amount of resistance in the coil. The coil resistance, reactance, and load determine the amount of current flowing through the coil. Current flow determines the strength of the pole's magnetic field.

ROTORS

One common type of rotor is called a squirrel cage. Instead of wires, copper bars are inserted into slots in the surface of the core. The ends of these copper bars are joined together, forming a series of closed loops arranged to form what looks like a cage.

The fields created by the stator electromagnets cut across the closed loops in the rotor and large currents are induced in the rotor loops. These induced currents create a magnetic field in the rotor with the opposite polarity of the stator electromagnet. Because opposite magnetic poles attract, the rotor is locked into a fixed position. The attraction of unlike poles results in a condition called locked rotor, meaning the rotor cannot turn. If the rotor is spun, it will continue to spin due to the alternating polarity of the stator poles.

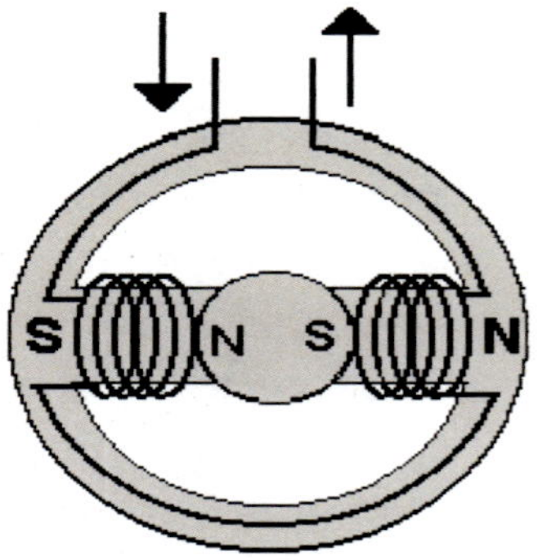

This push-pull action is continuous as the poles reverse polarity and the rotor tries to catch up with the changing polarity.

Fig. 3-5: Rotor polarity is opposite of stator poles

MOTOR SPEED

Motor speed is determined by the number of stator poles. Rotor speed is measured in revolutions per minute (RPM). Synchronous speed is determined by dividing 7,200; the number of alternations (changes from positive to negative) per minute in a 60 Hz circuit by the number of stator poles . A motor running at full load actually rotates at a speed about 4 to 5 percent below synchronous. This difference in motor speed is called slip.

MOTOR SPEEDS		
NUMBER OF POLES	SYNCHRONOUS SPEED	ACTUAL SPEED
Two-pole Motor	7200 ÷ 2 = 3600 RPM	3450 RPM
Four-pole Motor	7200 ÷ 4 = 1800 RPM	1725 RPM
Six-pole Motor	7200 ÷ 6 = 1200 RPM	1150 RPM

Fig. 3-6

SHADED POLE

Shaded pole motors are the simplest type of induction motors. All single-phase motors require a means of producing a second magnetic field for starting. In a shaded pole motor, the face of each stator pole carries a copper ring called a shading coil. Currents in this coil delay the magnetic flux in part of the pole to provide a second rotating field. This produces a low starting torque compared to other classes of single-phase motors. Shaded pole motors have only one winding and no capacitor or starting switch, making them economical and reliable. Their low starting torque is best suited to driving fans or other loads that are easily started and require under one-half horsepower. Shaded pole motors rotate toward the shading band.

The photo in Figure 3-7a shows a common C-frame motor. With the shading coils positioned as shown, this motor will start in a clockwise direction as viewed from the shaft end.

Fig 3-7a: Common C-frame motor

Fig 3-7b: Shaded pole motor

SPLIT-PHASE MOTORS

For higher starting torque, a start winding is required. The start winding establishes another magnetic field in the stator that is out of phase (out of step) with the main winding. These are called split-phase motors. The phase angle, or difference in magnetic force, of the start winding can be accomplished using different methods. For the split phase motor, the start winding placement relative to the run (main) winding, higher resistance and lower inductive reactance will give the motor starting torque.

Start windings are made with smaller diameter wire than run windings and have more turns on their laminated poles. This higher resistance in a start winding produces a magnetic field that lags the run winding.

At start-up, current flows through both windings. Magnetism in the start winding is slightly behind the run winding due to its higher resistance. These two magnetic fields, one behind the other, create the necessary turning force on the rotor.

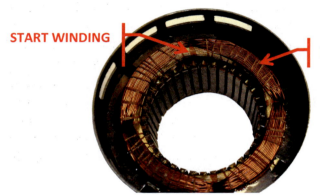

Fig. 3-8: The start winding creates a magnetic field that is out of phase with the run winding

DIRECTION OF ROTATION

Rotation, clockwise or counter-clockwise, is always determined by the direction of current flowing through the start winding. To reverse rotation, the two power supply connections to the start winding must be reversed. This reverses current flow through the start winding, causing opposite polarity and rotation.

In open motors, the electrical connections are located at one end of the motor, called lead end, and the motor shaft exits the opposite end, called the shaft end. Rotation direction is usually determined by viewing the shaft end. However, rotation for General Electric motors is determined by viewing the lead end.

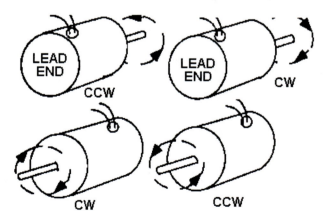

Fig. 3-9: Rotation of motor shaft depends on which end of the motor is being viewed

When recording rotation direction, always indicate which end of the motor is being viewed. If the shaft is turning to the right, rotation is clockwise (CW). If the shaft is turning to the left, rotation is counter clockwise (CCW). Not all motors have wiring connections that allow the motor to be reversed in the field.

DISCONNECTING THE START WINDING

The only purpose of the start winding in a split-phase motor is to start the motor. Once the motor has started, the start winding must be removed from the circuit or it will burn out. Fractional horsepower (Hp) motors use a centrifugal switch or relay for disconnecting the start winding after start-up. Open motors have a centrifugal switch inside the motor that is connected in series with the start winding.

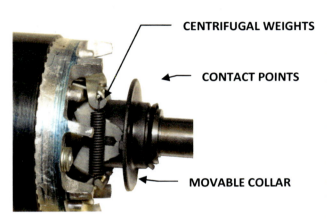

Fig. 3-10: Centrifugal switch

Motors of less than one horsepower are called fractional horsepower motors. For example, 1/2 hp and 1/4 hp are fractional horsepower motors.

The contacts on the centrifugal switch in Figure 3-10 are closed when the motor is not running. When the motor achieves 75 percent speed, enough centrifugal force is produced to make a pair of weights swing outward and open the switch. When the motor stops, springs pull the weights back and the switch is closed for the next start-up. Failure of the centrifugal switch to open permits the start winding to remain in the circuit, resulting in high amperage. In this case, the motor overload should stop the motor.

A current relay or positive temperature thermistor (PTC) can be used on fractional horsepower, open motors, or sealed refrigerant compressors to disconnect the start winding. The current relay coil produces a magnetic field from the high starting current, lifting the contacts closed and energizing the start winding. After the rotor starts turning, current decreases and the contacts open.

Fig. 3-11: Centrifugal switch contacts

Fig. 3-12: Current relay

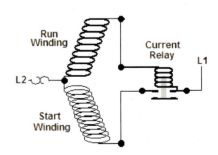

Fig. 3-13: Current relay wiring

A PTC is a thermistor with very low resistance at room temperature and very high resistance when hot. It is connected in series with a supply power and start winding. As current flows through the PTC, it heats up, creating a large voltage drop and removing nearly all current to the start winding.

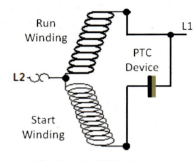

Fig. 3-14: PTC wiring

CS OR CSIR MOTORS

To improve starting torque, a start capacitor is connected to the start winding of an induction start motor. Single-phase induction motors with start capacitors are used when a motor must start under loaded conditions. Extra torque is required for starting purposes only. Once started, the capacitor and the start winding must be disconnected from the circuit. In HVACR systems, these motors are commonly found on belt drive fans, pumps, and fractional horsepower refrigeration compressors.

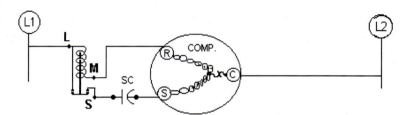

Fig. 3-15: CSIR motor wiring

PSC MOTORS

A permanent split capacitor (PSC) motor is often used to operate fans, blowers and refrigeration compressors. A PSC motor uses a run capacitor in series with a start winding while in operation. The run capacitor has two functions: keeping the start winding 90 degrees out of phase, and limiting current flow through the start winding to keep it energized. In this way, the start winding assists the run winding, making the motor operate more efficiently.

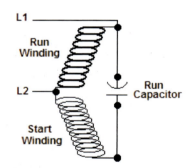

Run capacitors have low microfarad ratings and act as throttling devices to limit the number of electrons flowing through the start winding. Adding capacitance improves power factor and motor efficiency. The run capacitor is connected into the start circuit permanently, hence the name permanent split capacitor.

Fig. 3-16: PSC motor wiring

CSR MOTORS

Single-phase motors with run and start capacitors are called capacitor start-capacitor run (CSR) motors. CSR motors are used when a motor must start under loaded conditions and run with efficiency. They have the highest starting torque of all split-phase motors. Both the start and run capacitors are used to start the rotor turning. Once the motor has started, the start capacitor is disconnected from the circuit while the run capacitor remains connected.

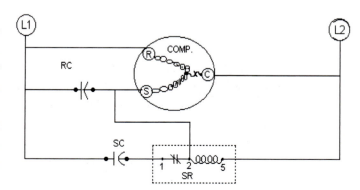

Fig. 3-17: Motor with potential relay wiring

A start capacitor can be removed from a start winding circuit by a centrifugal switch, current relay, PTC, potential relay (voltage relay), or an electronic timing device.

IDENTIFYING HERMETIC MOTOR TERMINALS

Technicians will often have to identify run (R), start (S), and common (C) terminals or check hermetic motor windings for an open, a short, or a grounded condition.

Note: When checking hermetic compressors, remember that terminals are under high pressure and could blow out of the shell, causing serious injury or death.

Step 1: Ohm each set of terminals with all wiring removed and record the information as shown in Figure 3-18(a).

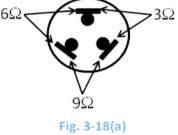

Fig. 3-18(a)

Step 2: The terminals with the highest ohm measurement (between run and start) identify the other terminal as the common, shown in Figure 3-18(b).

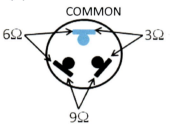

Fig. 3-18(b)

Step 3: Place one test lead of the ohmmeter on the common terminal and the other on the terminal with the lowest measurement. This is the run terminal. See Figure 3-18(c).

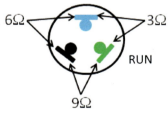

Fig. 3-18(c)

Step 4: The remaining terminal with the middle resistance measurement is the start terminal, shown in Fig. 3-18(d).

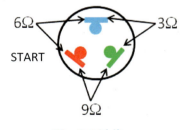

Fig. 3-18(d)

If any of the following conditions is found, the motor must be replaced:
- Any set of terminals measure 0 ohms (the winding is shorted)
- Any set of terminals measure infinity (the winding is open)
- Each terminal must be checked to ground. If a measureable resistance is detected, the motor is grounded.
- If the internal overload is open, there will be measureable resistance between start and run terminals and infinity between common to start or run. The compressor should be allowed time to cool and rechecked.

ELECTRONICALLY COMMUTATED MOTORS (ECMs)

ECMs run at very high efficiencies for single-phase motors. An ECM is a brushless DC motor operated by an attached variable frequency drive. The rotor has three permanent magnets mounted on the outside to create three poles. The stator has three windings, just as a three-phase motor does. An electronic inverter control mounted to the back of the motor pulses a DC voltage, to energize one winding at a time. The frequency drive can be set for the proper speed for heating and cooling or variable speed operation, depending on the manufacturer's controller.

Fig. 3-19: Electronically commutated motor

When used as a multi-speed motor, the air handler has a circuit board with dipswitches or wire jumpers that must be selected for proper cooling or heating CFM. Electronic thermostats monitor the indoor relative humidity as well as temperature, and vary blower speed to control humidity and CFM for stages of heating and cooling. Operation of the ECM on a blower is quite different than on a regular motor. When blower air flow is blocked, the motor current increases to maintain the correct RPM for the air flow setting, whereas in a PSC or other motor, current decreases with air flow. This inherent property of ECM blowers allows for operation at a higher duct static pressure. The motor saves energy by running at approximately 50 percent cooling speed when the thermostat fan switch is set to the "continuous" position.

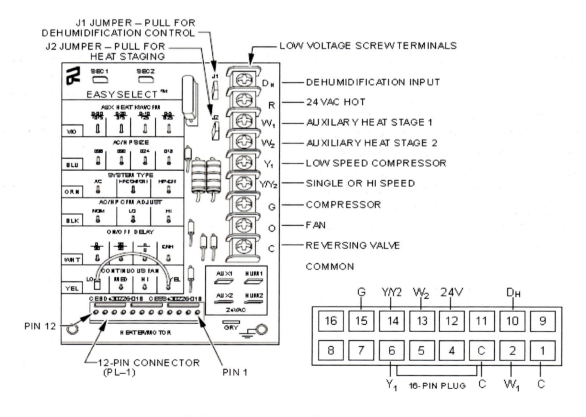

Fig. 3-20: ECM control board

SHADED POLE AND PSC MOTOR SPEEDS

Induction fractional horsepower motor speed is determined by the number of run windings added to the stator poles or the number of poles used. A multi-speed motor has several external taps (wires) for selection of motor speed. The desired motor speed tap is connected to the power source. The supply voltage (L1 & L2) must not be connected between two speed taps. The low resistance between speed taps will cause the motor to burn out as soon as power is applied.

CAPACITORS

Capacitors are often used to improve the operating characteristics of single-phase motors. A capacitor is an electrical device that stores and discharges electrical energy, causing more of a phase shift than a shading coil and winding does. Capacitors increase and decrease the magnetic field produced by the motor start winding. There are two different types of capacitors: start and run. All start and run capacitors are wired in series with start windings and are in parallel with each other on start-up.

Many capacitors have a bleed resistor soldered or connected to the terminals to safely discharge the capacitor each time the circuit opens. The purpose of the resistor is to reduce the severity of arcing that occurs at the relay contacts and reduce the possibility of a technician being shocked when removing a capacitor with a bleed resistor from the circuit.

Fig 3-21: Bleed resistor

START CAPACITORS

A start capacitor is connected in series with a motor start winding and switch. The applied voltage forces one side of the capacitor to fill with excess electrons while the other side discharges its excess electrons into the start winding. When the current alternates, the empty side again fills up while the other side discharges. The electrons rush into and out of each side of the capacitor, according to the alternating current. When one side is full, the other is empty. One side of the start capacitor discharges excess electrons into the start winding. This increases current flow in the start winding and increases the strength of the magnetic field, providing better starting torque. The start capacitor is not designed to stay in the circuit after the motor starts. It is typically energized for between .75 and one second and should never exceed four seconds. Start capacitors are easily damaged and can tolerate about twenty starts per hour without overheating. Start capacitors are usually constructed with a plastic bakelite case.

Replacement capacitors should have the same microfarad and voltage rating as the one being replaced. In an emergency situation, a replacement start capacitor up to 20 percent over capacity may be used as a temporary repair. Never install a capacitor that is under capacity.

Fig. 3-22: Start capacitor

THE RUN CAPACITOR

Run capacitors are made of dielectric material and conductors, enclosed in an oval or round metal case containing a dielectric oil to help dissipate heat. A run capacitor increases motor running torque by keeping the start winding slightly energized during the run cycle. Run capacitors are connected in series with start windings, just like start capacitors; however, run capacitors are not disconnected. They limit the amount of electrons entering the start winding and many are dual types marked with hermetic, fan, and a common center terminal.

A run capacitor usually has an identifying mark, like a minus sign or red dot, on one of the terminals, showing which plate is closest to the metal case. The hot wire should be wired to this terminal. If the capacitor becomes shorted to the case, the main fuse or circuit breaker will open. This will prevent the motor from operating without both of its windings, or ground current thru the start winding limiting the potential of overheating and damage.

Fig. 3-23: Run capacitor

CAPACITOR RATINGS

Capacitors have two important ratings. One is VAC (volts, alternating current) and the other is microfarads (mfd or µf). Microfarads indicate the capacitor's energy storage capacity. Start capacitors have much higher microfarad ratings than run capacitors. Run capacitors are approximately 1.5 mfd to 60 mfd, while start capacitors can be up to 1,600 mfd. Most start capacitors have a range of mfd ratings, such as 88 - 108 mfd.

When capacitors are connected in parallel, their microfarad ratings are added together. When connected in series, the microfarad rating is lower than the lowest capacitor rating.

The formula for finding total capacitance in series is:

$$C\ Total = \frac{(C1 \times C2)}{(C1 + C2)}$$

The peak voltage rating of a capacitor must be the same or higher than the OEM capacitor. The back EMF produced in the start winding while the motor is operating is higher than the applied voltage. If peak voltage is more than the rating of the capacitor, the capacitor plates will short or open. When capacitors are connected in parallel, applied voltage cannot exceed the lowest voltage rating. When capacitors are connected in series, applied voltage cannot exceed the sum voltage rating.

SINGLE-PHASE MOTOR STARTING RELAYS

Hermetic refrigeration compressors cannot use internal relays to disconnect start windings after start up. Depending on compressor size, there are several options for external motor starting relays.

CURRENT RELAY

The current relay is used on motors normally rated one horse power or less (See fig. 3-15). The coil of a current relay is made of 18- to 14-gauge copper wire for low resistance (usually less than one ohm). The main terminals of a current relay are L, M, and S, with L connecting to line, M connecting to run winding, and S connecting to the start winding or start capacitor. Since the coil is connected in series with the run winding, it must have a very low voltage drop and be able to carry the amperage draw of the motor. The contacts are normally open (NO) and the sequence of operation is:

Fig. 3-24: Current relay

Motor off—contacts open
Motor energized—contacts closed
Motor starting amperage drops—contacts open

Starting current (LRA) through the coil creates a magnetic field that pulls the solenoid up to close the contacts in series with the start winding or start capacitor. As the motor comes up to speed, amperage decreases, magnetic force reduces, and gravity pulls the solenoid down to open the start contacts.

To check the contacts, an ohmmeter can be connected to the L and S terminals. There should be continuity with the relay turned upside down. Due to movement and the way contacts close, the relay should be turned over three or four times and checked each time. Usually, a coil failure is easily detected by a visual inspection for a burnt spot and a check for continuity between L and M.

POTENTIAL RELAY

A potential or voltage relay has a coil and set of normally closed contacts wired in series. The main terminals of a potential relay are 1, 2, and 5. Terminal 1 connects to the start capacitor, terminal 2 to the start winding at the run capacitor, and terminal 5 to the common of the compressor (See fig. 3-17). Note: The leg of line voltage feeding a run winding must also feed the capacitors.

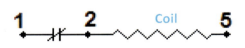

Fig 3-25: Potential relay wiring

Fig. 3-26: Potential relay

A relay coil is identified as terminals 2 and 5 and has a very high resistance. The normally closed (NC) contact is identified as terminals 2 and 1. The relay coil is wired parallel to the start winding. Contacts are wired in series with the start capacitor and the line feeding the run winding. The relay coil is energized by the counter (back) electromotive force from the start winding. Voltage induced in the start winding is greater than applied line voltage. Because the start winding has a higher resistance and more turns than the run winding, it acts like a step-up transformer. As the rotor begins to turn, magnetic lines of flux cut through the windings at a faster rate, causing the back EMF to increase with the rotor speed. As the compressor gets to 75 percent operating speed, voltage becomes high enough to energize the potential relay coil and the contact opens. The relay has three ratings: continuous coil voltage, contact pick-up voltage (open), and contact drop-out voltage (close).

SOLID STATE AND PTC
There are two solid-state or electronic starting devices: positive temperature co-efficient thermistors (PTCs) and electronic relays. Either of these can be wired in series with a start winding, with or without a start capacitor (See fig. 3-14).

A PTC has a resistance of approximately 5 to 15 ohms cold. As current passes through, the PTC heats up and resistance increases to 10,000 ohms or higher. This high resistance virtually stops current flow to the start winding and/or capacitor. A PTC can be checked by wiring it in series with a 100 watt, 120 Vac incandescent light bulb. When power is applied, the light should be bright, and then fade out in about 60 seconds.

Fig. 3-27: PTC relay with capacitor

An electronic start relay relies on either voltage sensing or time delay to disconnect the starting circuit. It is wired with only two leads in series with the start capacitor, connected to the run and start terminals of the motor.

CALCULATING MOTOR HORSEPOWER
Motors are rated according to the amount of torque they can produce. This turning power is measured in horsepower. The electrical consumption is measured in watts. A one horsepower motor consumes approximately 746 watts. Thus, a five horsepower motor will use approximately 3,730 watts (5 X 746=3730). This formula assumes 100 percent efficiency. When wattage and voltage are known, Ohm's Law allows us to determine amperage draw at full load conditions. True horsepower must take into account the power factor, which is approximately 0.95, motor slippage and friction. Field calculated wattage will be higher as much as 30% or more depending on type of motor.

SERVICE FACTOR
Service factor (SF) is a multiplier marked on data plates that indicates the total permissible horsepower loading when the motor is operating at rated voltage and frequency.

Example:
A motor rated at 0.75 Hp with a service factor of 1.2 is rated at 0.9 Hp.
(0.75 HP X 1.2 SF = 0.9Hp)

TIP: When replacing a motor, the replacement motor's horsepower multiplied by its service factor must be equal to or greater than the original motor's horsepower multiplied by service factor.

LOCKED ROTOR AMPS (LRA)
At start-up, before a rotor starts to rotate or if it is locked, current flow in the motor is determined by the resistance of the windings. Starting current (inrush current) is about six times higher than normal running amperage. This high current flow is called locked rotor amps (LRA). As the motor picks up speed, counter-EMF is generated and reduces current flow. At operational speed, current flow is determined by resistance of the run winding, counter-EMF generated by the motor, and the load placed on the motor.

FULL LOAD AMPS (FLA)

Full load amps (FLA) or rated run load amps (RLA) refers to the amperage a motor draws when at normal speed and fully loaded. Most induction motors operate at less than FLA because the motor is rarely working at fully loaded conditions. Overload occurs when amperage exceeds a percentage of the FLA rating or motor heat gets to high.

OVERLOAD PROTECTORS

An overload protector is a device that protects a motor against overload conditions. There are a variety of overload protector types, and the type commonly used depends on the type of motor and its application. It is common for a fractional horsepower, single-phase AC motor to have an overload with a snap-acting bimetal disc that makes and breaks a set of contacts. Excessive current causes the bimetal disk to deflect and open the circuit. When cooled, the disc returns to the closed position.

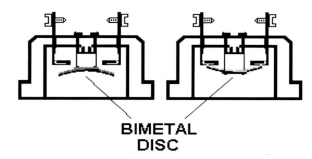

Fig. 3-28: Overload protectors use a snap-acting bimetal disc to break a set of contacts

Repeated opening and closing of the overload is called cycling. Continuous overload cycling is a warning that the motor is in danger of a burnout. It is important to isolate and repair the cause of the overload to prevent damage to the motor.

Another type of overload protector is an internal overload protector located inside the motor windings. When checking motor windings, do not condemn the motor until allowing sufficient time for the internal overload to reset. The overload is activated by temperature and current and may take four to eight hours to reset.

 When servicing a motor that has tripped on overload, be sure to disconnect it from the power source. The motor could suddenly restart when the overload resets.

THREE-PHASE MOTORS

Three-phase motors are very common in commercial and industrial applications. They are smaller and more efficient than single-phase motors of equal horsepower. Three-phase motors have high starting torque and high running torque, without the use of start windings or capacitors.

For proper operation, all three supply wires (L1, L2, and L3) must be connected to the motor terminals. The safety ground (green) wire is included for equipment ground. This grounding wire is connected to the motor frame to provide an escape for electrons in case the motor windings become shorted to the metal frame.

Three-phase motors have three pairs of stator poles; one pair for each supply wire. Each winding produces a north and south pole; this is called one pole per phase. A typical three-phase motor has three pairs of stator poles, meaning three north and three south poles. Each north and south combination is located directly opposite another. These poles are equally spaced in a circle, exactly 60 degrees apart.

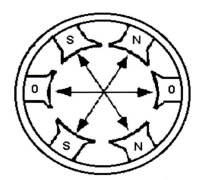

Fig. 3-29

The resistance of all three windings in a three-phase motor is the same. For single voltage motors, only one end of each winding is brought outside the motor for connection to the power source. The other end of each motor winding is factory connected inside the motor.

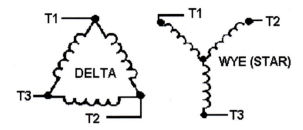

Fig. 3-30: Schematic symbols for Delta and Wye three-phase

With three-phase alternating current, the three power supply wires are 120 degrees out of phase and take turns changing polarity from north to south. When one pole is north, the second is rising and the third is dropping. This changing of polarity in the supply wires produces a strong rotating magnetic field in the stator poles that are out of step with each other. The alternating current produces different levels of magnetism in the set of stator poles producing the rotating push-pull effect on the rotor. In Fig 3-31 the six physical poles are separated by 60 electrical degrees.

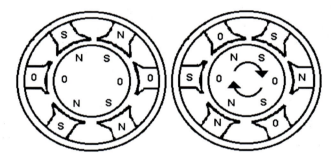

Fig. 3-31: Stator polarity produces a push-pull effect on the rotor

CHANGING THE ROTATION OF A THREE-PHASE MOTOR

Direction of rotation is determined by the direction of the rotating field. Reversing rotation on a three-phase motor is done by interchanging any two supply wires. This simple procedure causes the magnetic field to rotate in the opposite direction.

 Improper rotation can be devastating to equipment and personal safety. If necessary, disconnect equipment before checking proper motor rotation.

CHECKING RESISTANCE OF WINDINGS

The motor windings on a three-phase motor can be checked with an ohmmeter. The motor must be disconnected from the circuit and the measurements obtained from one motor lead to another. If a resistance measurement of zero is obtained, the winding is shorted. If a measurement is obtained from any motor lead to ground, the winding is grounded. A measurement of infinite resistance indicates that the winding is open. In each of these cases, the motor must be re-wound or replaced.

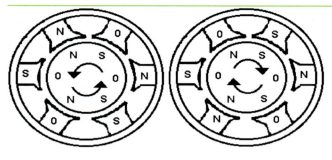

Fig. 3-32: Changing any two supply wires changes the rotation of the rotor

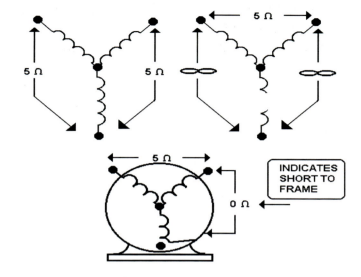

Fig. 3-33: An ohmmeter is used to check motor windings

The resistance measurement on three-phase motors windings will vary from less than one ohm to 50 ohms, depending on motor size. The larger the motor, the lower the resistance. Each winding has the same resistance, except for dual-voltage windings. A dual-voltage motor has its main windings split into six windings of equal resistance.

DUAL-VOLTAGE THREE-PHASE MOTORS

Many three-phase motors are designed for connection to either of two different voltages: 240 or 440. These are called dual-voltage motors. Instead of having just three external wires to connect, these motors have nine or twelve. The wires are tagged and numbered for easy identification. NEVER remove these numbers.

Regardless of which voltage is being connected, all wires must be connected properly. When connected to the lower voltage, the windings are connected in parallel; when connected to the higher voltage, they are connected in series. The three power supply wires are ALWAYS connected to motor numbers T1, T2, and T3.

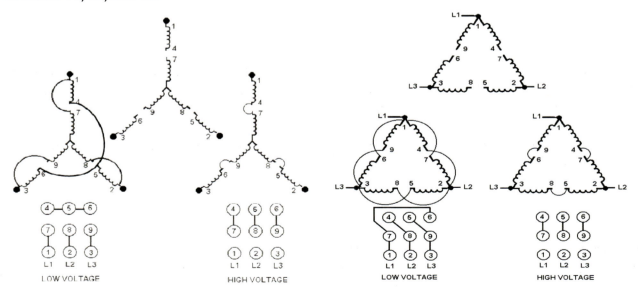

Fig. 3-34: Low- and high-voltage connections for a WYE (Star)-connected motor

Fig. 3-35: Low- and high-voltage connections for a Delta-connected motor

THE MOTOR NAME PLATE

Instructions for making electrical connections to a motor are usually included on the motor nameplate, also called the data plate. The nameplate should be carefully viewed before selecting, replacing, or wiring a motor. Figures 3-36 and 3-37 show an example of a single-phase motor nameplate and a three-phase motor nameplate.

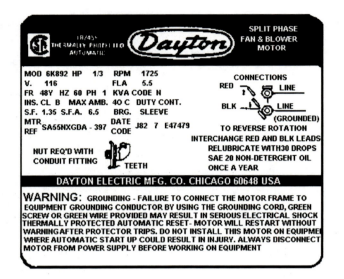

Fig. 3-36: Single-phase motor nameplate

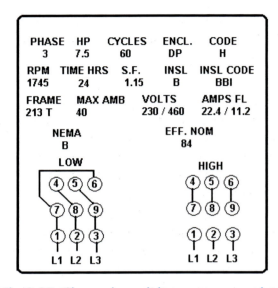

Fig. 3-37: Three-phase delta motor nameplate

NAME PLATE DATA DEFINITIONS

Frames and Type: Motors of a given horsepower rating are built in a certain size of frame or housing. NEMA has standardized the frame size and shaft heights to be used for each integral horsepower motor. This permits easy replacement or interchanging of motors.

Max Amb: The maximum ambient temperature at which a motor can be operated

Temperature Rise: The amount of temperature rise permitted above ambient air at rated load

Duty (Time / Hours): All electric motors are designed for either continuous or limited duty. Those designed for continuous duty deliver the rated horsepower for an indefinite period without overheating. Limited duty motors deliver rated horsepower for a specified period of time and will overheat if operation is extended. Limited duty motors are often used to operate valves, pumps, or louvers.

Thermal Protection: Indicates type of thermal protection provided, if any

FLA (Amps FL): The rated current (in amperes) when operated at full load

LRA (Locked Rotor Amps): The rated current (in amperes) if rotor is unable to turn

KVA Code: A letter indicating the starting current required; the higher the locked-rotor-kilovolt-ampere (kVa), the higher the starting current surge

Insulation Class (INSL): A designation for the type of insulation used; primarily used for rewinding purposes

Service Factor (S.F.): The amount of overload that a motor can tolerate on a continuous basis at rated voltage and frequency.
Example:
A motor rated at 0.75 Hp with a service factor of 1.2 is rated at 0.9 Hp.
(0.75 HP X 1.2 SF = 0.9Hp)

VARIABLE FREQUENCY DRIVES (VFDs)

Variable frequency drives are used to vary the speed, torque, horsepower, and direction of single and three-phase AC motors. The input AC voltage passes through a filtered rectification circuit to change it to direct current. By using electronic switches, the frequency can be varied from approximately 1 to 400 hertz. Today's VFD's have programmable options which can be programmed changing for different applications. When changing the voltage and the frequency at the same time, a motor will produce approximately 150% of torque with only 50% of current. The controller can be programmed to slowly ramp the motor up to speed and gradually slow it down before stopping. This eliminates the high mechanical stress developed from starting and stopping. When an AC motor needs to be prevented from freewheeling or needs to come to a stop very quickly, a low DC current can be applied to the windings, which will act as a break. Infinite capacity control may be obtained when used in conjunction with air conditioning or refrigeration compressors. For other applications, minimum and or maximum speeds can be set.

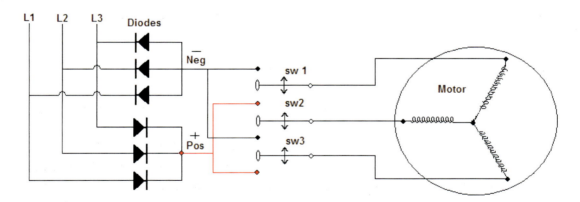

Fig. 3-38: Simple VFD diagram

VARIABLE-SPEED DRIVES (VSDs)

Variable speed drives can be used on DC (direct current), single-phase and three-phase AC motors. They differ from variable frequency drives in switching action. The principle mode of operation is pulse with modulation. The amount of time the switches are opened and closed is used to vary motor speed. The controls can range in voltages from 110 to 10Kv.

Student Worksheet

Page 73

Chapter 3: Motors

Name _____   Date _____

1. Test several compressors assigned by the instructor for a ground.
2. Test a compressor assigned by the instructor for shorted windings.
3. Find the resistance and identify the common, run, and start terminals of four different compressors assigned by instructor.

Compressor 1

A. _____

B. _____

C. _____

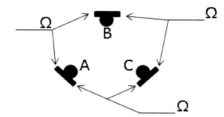

Compressor 2

A. _____

B. _____

C. _____

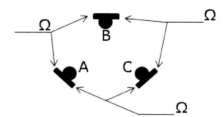

Compressor 3

A. _____

B. _____

C. _____

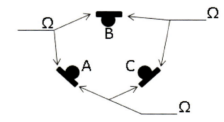

Compressor 4

A. _____

B. _____

C. _____

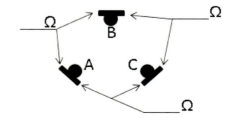

Electrical Theory & Applications for HVACR   ©2012 ESCO Group

Student Worksheet

Chapter 3: Motors

Name _____   Date _____

1. Test a start capacitor with an ohmmeter for an open or short.

2. Test a run capacitor with an ohmmeter for an open or short.

3. Use a capacitor analyzer to check the micro-farad rating of a start capacitor.

4. Use a capacitor analyzer to check the micro-farad rating of a run capacitor.

5. Use a capacitor analyzer to check the micro-farad rating of a dual run capacitor.

Student Worksheet

Chapter 3: Motors

Name _____   Date _____

1. Check a current relay coil and NO contacts using an ohmmeter.

   Coil   _____

   Contacts   _____

2. Check the coils and contacts of three potential relays assigned by instructor.

   Relay 1:   Coil: _____ ohms   Terminal numbers: _____

                Contact: _____ ohms

   Relay 2:   Coil _____ ohms   Terminal numbers: _____

                Contact _____ ohms

   Relay 3:   Coil: _____ ohms   Terminal numbers: _____

                Contact: _____ ohms

Chapter 4: Understanding Wiring Diagrams

OBJECTIVES:

- Define the different types of diagrams
- Dissect and interpret various wiring diagrams
- Define the symbols used for loads, switches, and safety devices
- Explain the operation of how and when loads are energized or de-energized
- Explain how switches are energized or de-energized
- Facilitate troubleshooting HVACR equipment
- Read and draw wiring diagrams

4 Understanding Wiring Diagrams

PICTORIAL AND SCHEMATIC DIAGRAMS

A wiring diagram is a simplified pictorial representation of an electrical circuit. Different types of diagrams have different applications in the HVACR industry. For example, installation diagrams are used by installers in making electrical connections for high and low voltages, but are not useful for technicians trying to troubleshoot equipment. Pictorial diagrams readily identify components and the wiring between them in detailed physical appearance and are used for locating electrical components, line connections, and wire colors, and for troubleshooting. Pictorials show how components and switches are actually wired, making simple circuits easy to understand. When many electrical components are involved, however, a pictorial diagram becomes too cumbersome to be used for tracing circuits or troubleshooting.

A schematic does not illustrate where components are located, but how they are connected in the circuit. Schematics are less cluttered than pictorials; they present the sequence of operation in an organized manner and use symbols to illustrate components. Experienced technicians prefer schematics because they make circuits easily traceable, facilitating troubleshooting.

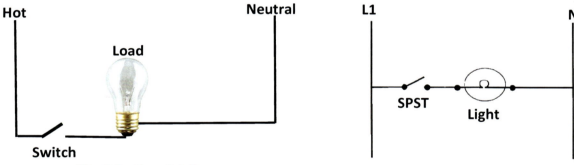

Fig. 4-1: Pictorial diagram

Fig. 4-2: Schematic diagram

Pictorials Versus Schematics

Pictorial drawings show components and wiring as they actually appear, whereas schematics use symbols to represent components and may not show specific locations. Figures 4-3 (a and b) each show a circuit represented by both a pictorial and a schematic diagram.

Combining Pictorial and Schematic Drawing

Figure 4-4 is a diagram for a Frigidaire© side-by-side refrigerator/freezer. It shows both a pictorial diagram for locating electrical components and a schematic diagram to clearly illustrate the various electrical circuits. Dual diagrams like this one are very useful when troubleshooting equipment with concealed wiring.

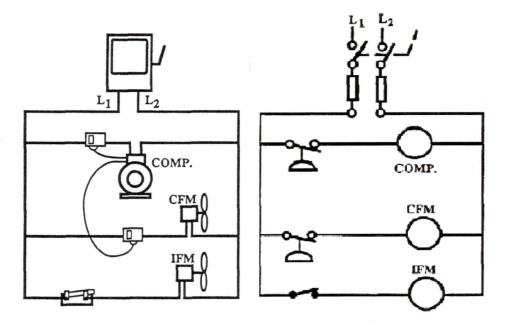

Fig. 4-3(a): Pictorial (left) and schematic diagrams of the same circuit

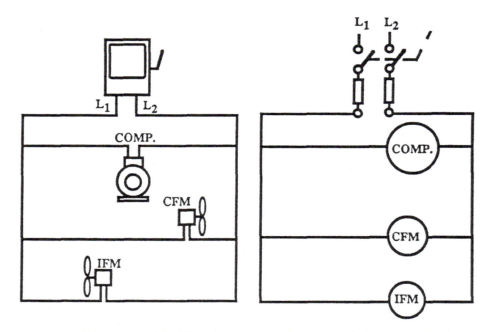

Fig. 4-3(b): Pictorial (left) and schematic diagrams of the same circuit

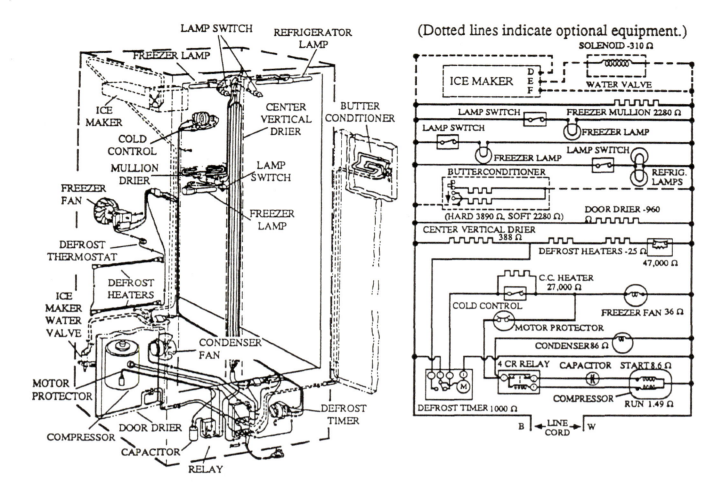

Fig. 4-4: Pictorial and schematic diagram for Frigidare© side by side.

LINE-SIDE VERSUS LOAD-SIDE

Switches located between the line leg and the load are called line-side switches. Switches located between the load and the load leg are called load-side switches.

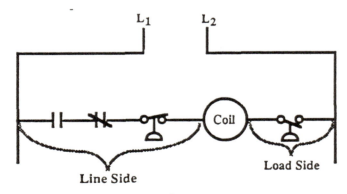

Fig. 4-5

LADDER DIAGRAMS

Another type of diagram is the ladder diagram. A ladder diagram (Figure 4-6) is arranged with the power supply lines drawn vertical as the legs of a ladder. Each horizontal line, or rung of the ladder, contains one load and its control switches. The load lines may be numbered for ease of identification. Control switches are usually located on the line side of the device being controlled; however, some manufacturers place a switch or two on the load side of the rung.

Note: Contacts and switches are always shown in the de-energized position on most diagrams, there are times where the contacts on switches are shown based on equipment operating cycle (example: ice machine harvest cycle).

To use a schematic or ladder diagram, a technician must have knowledge of equipment components and their corresponding electrical symbols. Most schematic diagrams contain a legend that lists any abbreviations and special components, as shown in Figure 4-6.

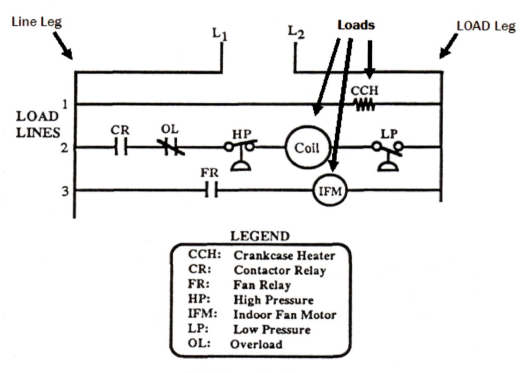

Fig. 4-6: Ladder diagram

READING A WIRING SCHEMATIC

Reading a wiring schematic can be made easier by following a few simple rules:

- Schematics are read like books, from top to bottom and left to right.
- There must be a complete circuit in order for current to flow through a component.
- Electrical contacts and switches are shown in their normal "off" position, unless otherwise stated on the diagram.
- When a relay is energized, all of its contacts change position. Normally open contacts close and normally closed contacts open. All contact symbols with the same coil number or letter are controlled by that coil, regardless of location in the circuit.
- Switches or contacts used to provide the stop function are normally closed and generally wired in series.
- Switches or contacts used to provide start function are normally open.

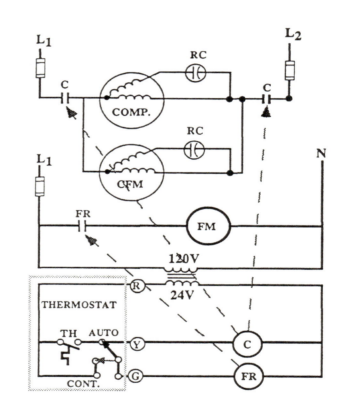

LEGEND

C: Contactor Coil
Comp: Compressor
CFM: Condenser Fan Motor
RC: Run Capacitor
FM: Fan Motor
FR: Fan Relay Coil
TH: Thermostat

Fig. 4-9: Wiring schematic

Schematic Diagram Symbols

Figure 4-10 shows some of the symbols commonly used on schematic diagrams.

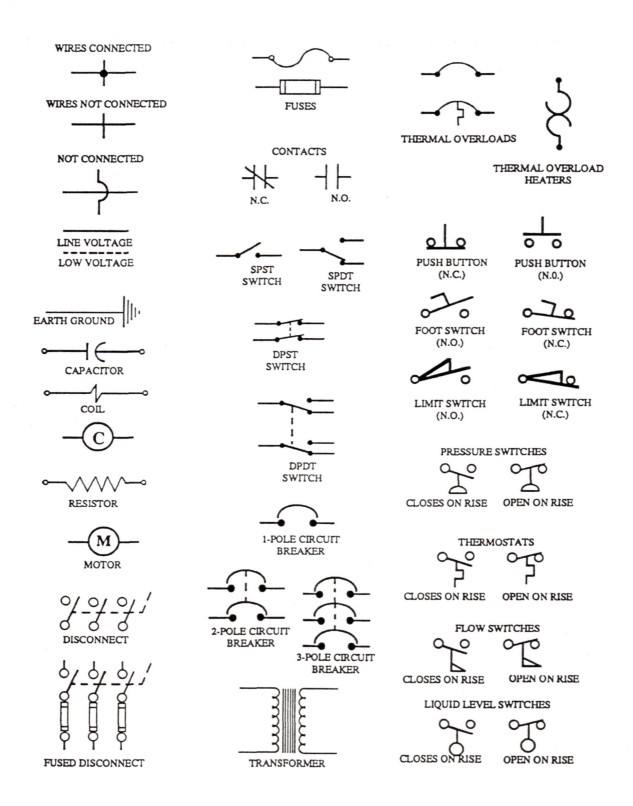

Fig. 4-10: Common schematic symbols

Student Worksheet

Page 87

Chapter 4: Understanding Wiring Diagrams

Name _____   Date _____

Typical 120 volt window air conditioner wiring schematic. Both motors are PSC's and the compressor requires a 3 to 5 minute delay before restarting.

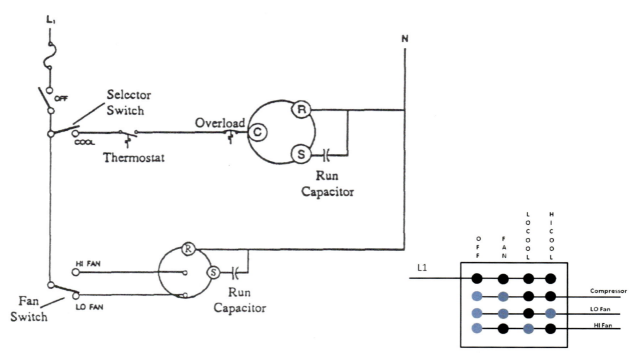

Black dots indicate electrical connections for switch position to L1.
Blue dots indicate no electrical connection to L1.

For each problem all electrical components are to be checked for defects:

1. With the system turned on, only the high speed fan runs.

2. Fan motor runs on either speed, but compressor does not run.

3. Fuse is good, but compressor and fan will not run.

4. On low cool, only the compressor runs.

5. On high cool, only the fan motor runs.

Student Worksheet

Chapter 4: Understanding Wiring Diagrams

Name _____   Date _____

1. How many line voltage "loads" are shown in the diagram?

2. How many low voltage "loads" are shown in the diagram?

3. What relay controls the condenser fan motor?

4. What switches must be closed to energize the control relay coil?

5. How many sets of contacts are controlled by the control relay coil?

6. With the "FAN" switch in the "CONT" position, what effect will closing of the thermostat have on the indoor fan relay?

7. When does current flow through the transformer primary?

8. With the "FAN" switch in the "AUTO" position, will the indoor fan start before the condenser fan motor?

9. How is the "COOL" switch operated?

10. Not counting the fused disconnect, how many contacts are in the compressor contactor coil load line?

Electrical Theory & Applications for HVACR   ©2012 ESCO Group

Student Worksheet

Page 91

Chapter 4: Understanding Wiring Diagrams

Name _____   Date _____

With the system in the normal operating mode, what voltages would you measure across the following?

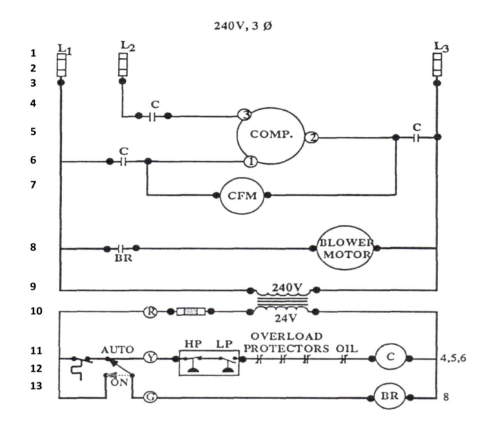

1. L1 to L2: _____

2. Across the coil in line 11: _____

3. R to G on the Thermostat: _____

4. Condenser Fan Motor: _____

5. Transformer Primary: _____

6. Across contacts on line 8: _____

7. From y to Line side of coil on line 11: _____

8. Across the fuse in line 10: _____

Electrical Theory & Applications for HVACR   ©2012 ESCO Group

List all electrical components that should be checked for possible causes in each of the following:

9. The thermostat contacts are closed, the indoor blower motor is running, but condenser motor and compressor are off.

10. The thermostat contacts are closed, the compressor and condenser fan motor are short cycling, but the indoor blower motor is off.

11. If the L3 fuse opened, would the condenser fan motor continue to run?

12. If the L3 fuse opened, would the evaporator fan motor continue to run?

Student Worksheet
Page 93

Chapter 4: Understanding Wiring Diagrams

Name _____   Date _____

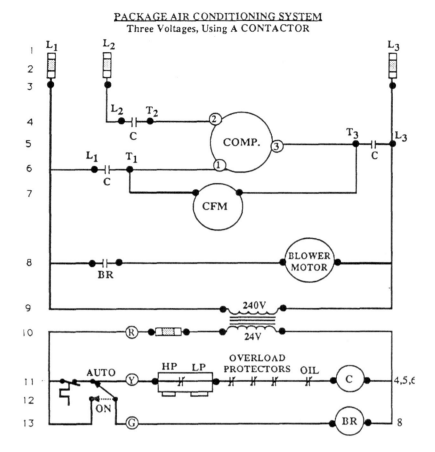

Draw lines to connect components below by using the schematic above.

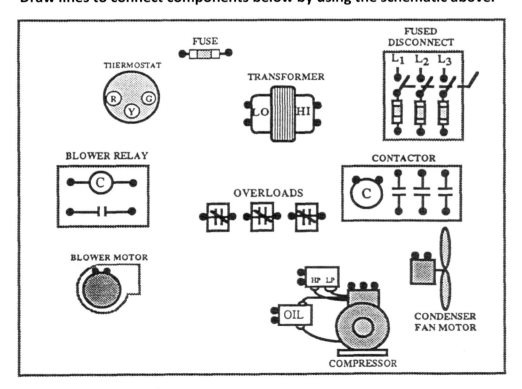

Student Worksheet

Page 95

Chapter 4: Understanding Wiring Diagrams

Name _____   Date _____

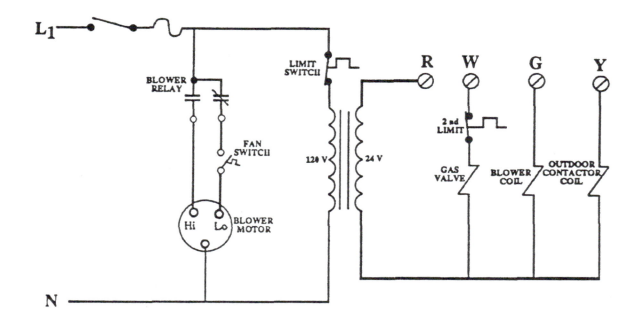

Draw lines to connect components below by using the schematic above.

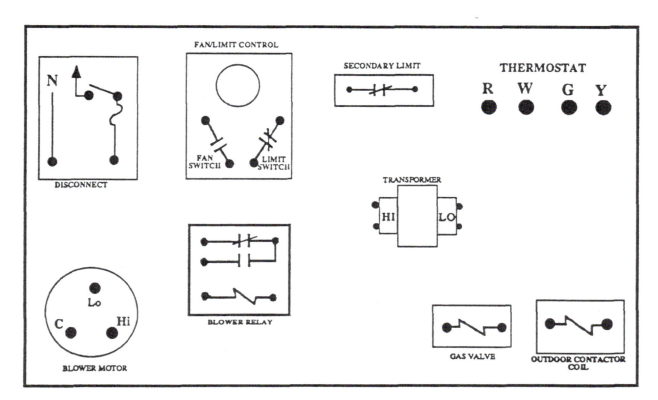

Electrical Theory & Applications for HVACR   ©2012 ESCO Group

Chapter 5: Automated Control Systems

OBJECTIVES:

- Understand basic concepts of automated controls protocols
- Connect and program a Wi-Fi enabled thermostat
- Program a programmable thermostat

Automated Control Systems

This chapter provides an overview of control standards and the various types of control systems available to homeowners.

WHY HAVE AUTOMATED CONTROLS

Due to the need to reduce energy consumption, controls used in the HVACR field are constantly evolving. Electronic processor technology has made it possible to turn analog household thermostats into microcomputers. With this technology, the ability to control comfort conditions has expanded to home computers and smart phones. Smart thermostats are energy management systems that can be connected to a home's network via cable or wireless internet, giving homeowners the ability to control their HVAC systems from anywhere with internet access.

It is vital for today's technicians to be computer literate and to have a basic understanding of electronics, integrated circuits, microprocessors, digital controls, wireless controls, and network communication systems.

Modern HVAC controls encompass much more than just a heat/cool thermostat; they allow homeowners greater flexibility, comfort, and energy efficiency. New controls can be used with simple heating and air conditioning systems, or with multilevel buildings that have hundreds of processors. In these complex systems, several processes and sequences are often occurring at once. Modern control systems allow these processes to be automated and controlled with a precision that was unavailable in the past. Building systems can be checked at a glance, and some systems can notify the technician of existing problems. If a problem is detected, it may be possible to temporarily resolve the issue by overriding a fault from a remote place. Time and money are saved when automated systems report small problems before they become large ones.

BUILDING AUTOMATION

Building automation refers to an intelligent network of electronic devices designed to monitor and control the mechanical, electronic, and lighting systems in a building. A building automation system (BAS) is an example of a distributed control system.

BAS can perform many vital core functions such as:
- Keeping the building climate within a specified range.
- Provide lighting based on an occupancy schedule.
- Monitors HVAC/R system's performance and device failures.
- Provides email and/or text notifications to the proper person.

BAS functionality reduces building energy and maintenance costs. A building controlled by a BAS is often referred to as an intelligent building system or a smart home.

SYSTEM PROTOCOLS

Most companies have their own proprietary controllers for specific applications. Some controls are designed with limited functionality, such as starting or stopping an HVAC system. Others are more flexible, allowing temperature adjustments and more. Most have proprietary software that works with ASHRAE's open protocol BACnet or open protocol LonTalk.

Communication standards are being developed that will allow all controls regardless of manufacturer to interface with each other. This will make it possible to improve functionality at a future time, without replacing entire control systems. The following is a list of the various networks and communication protocols in use today.

Consumer Electronic Bus (CEBus): A protocol designed to be used over the power lines by utility companies. This system was intended for use with smart type electrical meters to monitor energy use.

LonTalk: Competed with CEBus by using power line technology. It is well established in the industrial space in parts of Europe and Asia.

Universal Plug and Play: Microsoft worked with the Home Automation & Security committee on this project. It is used for connecting Web-enabled devices to computer systems.

Simple Control Protocol: Was developed by GE, CEBus and Microsoft. This protocol automatically allowed a computer to recognize the enabled device when plugged into the power outlet.

ZigBee: Uses Web Service Device (WSD) a protocol, which interfaces with a device, allowing it to be seen and controlled using third-party software.

BACnet: A data communication protocol for building automation and control networks. BACnet was developed by a committee formed by the American Society of Heating, Refrigerating and Air-Conditioning Engineers (ASHRAE). The main set of rules govern the exchange of data over a computer network. It covers the type of cable used and how to form a particular request or command from the communications equipment.

Climate Talk Alliance: An organization of companies committed to developing a common communication infrastructure for HVAC and Smart Grid devices. Their goal is to enable diverse systems and organizations to work together.

BASIC OPERATING STRUCTURE

The operating technology of automated controls uses analog and digital signals. Analog signals are used as input or output for variable measurements or control of temperatures, motor speeds, and pressures.

Digital signals are used to turn things on or off and to indicate "on" or "off." In either case, the control signal is normally low-voltage and, depending on the control, AC or DC.

CONTROLLER TYPES

As with all standards, there are different controllers used to adapt system components to the BAS. The three main types are programmable logic controllers (PLCs), system/network controllers, and terminal unit controllers.

NETWORK CONNECTIONS

Remote connections to most systems now available require the modem IP address. Most internet providers connect homes with a dynamic connection in which the IP address will change. If a static IP address is not available, a Domain Name Service (DNS) must be used to keep track of the IP address in use.

The following is a graphic representation of the components of a home network with the HVAC system connected. HVAC controls can be hard wired or connected wirelessly to the home network.

Student Worksheet

Chapter 5: Automated Control Systems

Name _____   Date _____

1. What has made it possible to replace old analog thermostats with microcomputers?
    A. Consumer demand
    B. Government demand
    C. Electronic processor technology
    D. Lower manufacturing and installation costs

2. A building controlled by (BAS) is often referred to as:
    A. An intelligent building system or a smart home
    B. A best asset system
    C. An intellectual building system
    D. A Beta Alpha System

3. Most companies have their own proprietary controllers for specific applications. Some controls are designed with limited functionality, such as:
    A. Mobile limitation
    B. Starting or stopping devices
    C. Computer limitations
    D. Determining proper design temperature

4. LonTalk competed with CEBus by using:
    A. Microsoft
    B. Mobile devices
    C. Satellite communication
    D. Power line technology

5. A DNS server must be used with a Wi-Fi connected control that uses a/an:
    A. Dynamic IP address
    B. Static IP address
    C. Floating IP address
    D. IPV6-SSID address

Chapter 6: Troubleshooting

OBJECTIVES:

- Understand basic terms
- Use different meters in troubleshooting switches and loads
- Troubleshoot simple and complex circuits and determine problems and solutions

6 Troubleshooting

The majority of service calls are electrical in nature; however, mechanical problems often become evident after a reported electrical problem is investigated. Troubleshooting electrical problems is one of the most common duties of an HVACR technician.

It is impossible to cover every possible problem in a textbook; a technician must rely on a solid foundation in the basic theories of electricity, develop the ability to read schematics and manufacturer's bulletins, understand the controls in a system, and possess a working knowledge of testing equipment. Making unsystematic measurements or randomly replacing parts rarely solves problems.

 It is important for a technician to develop a systematic and methodical procedure for troubleshooting. This can only be accomplished by having a full understanding of electrical fundamentals, Ohm's Law, the rules for series and parallel circuits, and related components.

Most HVACR electrical systems consist of loads and control devices in parallel circuits with safeties, wired in series to complete or interrupt the electrical path when a limit is exceeded. Controls regulate normal functions and limits of a system. Safety controls are used to shut down a unit if the system experiences mechanical difficulties.

Components can be classified as either current-passing or power-using. In a very simple circuit, a switch is the current-passing component and a light bulb is the power-using component (load). In an HVAC-R circuit, the thermostat is generally a current-passing component and the relay coils, contactor coils, fan, and compressor motors are power-using components.

TERMS

The terms *primary* and *secondary* may be used to describe controls. A primary control is one that is used turn a load on or off for normal operation; a wall thermostat, fan relay, contactor are primary controls. A secondary controls are temperature and pressure switches, or other components used for safety.

Primary and *secondary* is also used to describe the two sides of a transformer. Primary is the input and secondary would be the output or load side.

Technicians must be familiar with the use of basic test equipment. The operation of voltmeters, ammeters, and ohmmeters has been previously discussed. This chapter applies that knowledge to circuit troubleshooting.

VOLTMETERS

A voltmeter measures the potential difference between any two points, much like a pressure gauge indicates pressure difference.

In the circuit shown in Figure 6-1(a), a measurement of 120V should be obtained if the voltmeter leads are placed between L1 and N. This shows that there is potential between the two measured points. However, the power-using device (the light bulb) is not operating because the switch is open.

In Figure 6-1(b), the voltmeter leads are still measuring the potential between L1 and N, but the power-using device is on.

Figure 6-1(c) illustrates that switches and fuses do not, or at least should not, use power. The potential between the two measured points is exactly the same, so there is no voltage measurement on the meter. If the contacts in the switch were corroded, the resistance would cause power consumption (heat) and a reading on the voltmeter. When a voltmeter is placed across a switch, any reading above 0.02 volts in a 120-volt circuit may indicate a failed component.

In Figure 6-1(d), the voltmeter leads are still measuring the potential between L1 and N, but the power-using device is burned open.

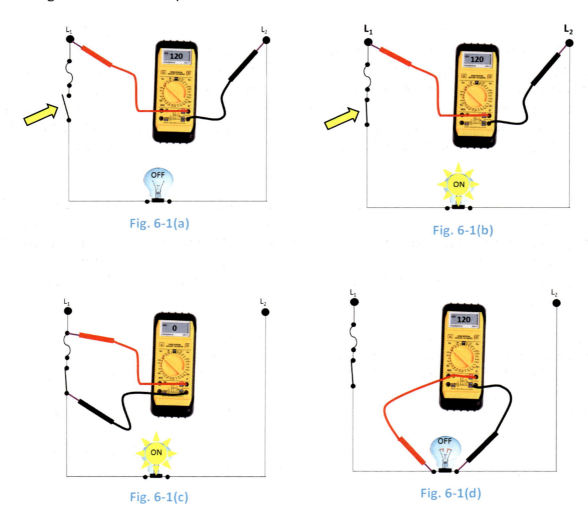

Fig. 6-1(a)

Fig. 6-1(b)

Fig. 6-1(c)

Fig. 6-1(d)

Most HVAC circuits contain more than one switch to control a device. The example in Figure 6-2 shows that a thermostat, high-pressure switch, low-pressure switch, and an overload (safety) must be closed to complete the circuit to the motor. There are two different ways of using a voltmeter to troubleshoot this circuit. Each technician can adopt the method that they are most comfortable with.

The sequence of troubleshooting should be: source, load, path.

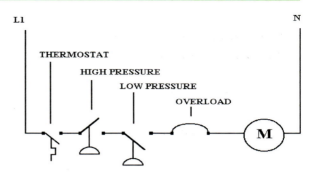

Fig. 6-2: Multiple switches/safeties

Lineal Searching

In a lineal search, start at the beginning of the circuit by testing for voltage between L1 and N. Then move the test lead from L1 to test for voltage at the thermostat, then the high-pressure switch, low-pressure switch, overload, and finally, the motor terminals. By checking each side of the switched components, open or faulty components or wiring can be found. This method locates the problem component, but can be very time consuming.

Split Searching

A split search includes some advanced steps. Check for voltage at the load first. If there is voltage at the load device, it is usually faulty and the rest of the circuit can be eliminated. If there is no voltage, choose a point about midway through the circuit, starting at the low-pressure switch. If there is voltage available there, eliminate L1, the thermostat, and the high-pressure switch as causes of the problem. Then measure the remaining half of the circuit until the problem is located. If no voltage was found at the midpoint of the circuit (low-pressure switch) the problem is before the test point. Split searching can make troubleshooting complex circuits easier and less time consuming.

TROUBLESHOOTING USING A VOLTMETER

The majority of service problems are electrical problems, which usually cause mechanical problems. Figure 6-3 is a 230-volt, single-phase electrical schematic of a typical commercial refrigeration system. The diagram includes a timer assembly with a defrost termination solenoid (DTS), evaporator fans, defrost heaters, temperature activated defrost termination/fan delay (DTFD) switch, low-pressure control (LPC), high-pressure control (HPC), compressor contactor assembly, and compressor/potential relay assembly.

The system is drawn in refrigeration mode and shows what voltages would be measured across certain points of the schematic if a voltmeter were used in troubleshooting. The diagram also shows where Line 1 (L1) is in relation to Line 2 (L2), for an easier understanding of the measured voltages.

Notice that anytime voltmeter probes see both L1 and L2, 230 volts is shown on the voltmeter. Anytime voltmeter probes see the same line (L1 to L1 or L2 to L2,) 0 volts are read on the voltmeter because there is no voltage difference between these measured points. So if the service technician can find where L1 and L2 are when troubleshooting, the rest is easy to determine.

Troubleshooting using a voltmeter

This diagram indicates the voltage measurements across the loads, switches and connections and ground.

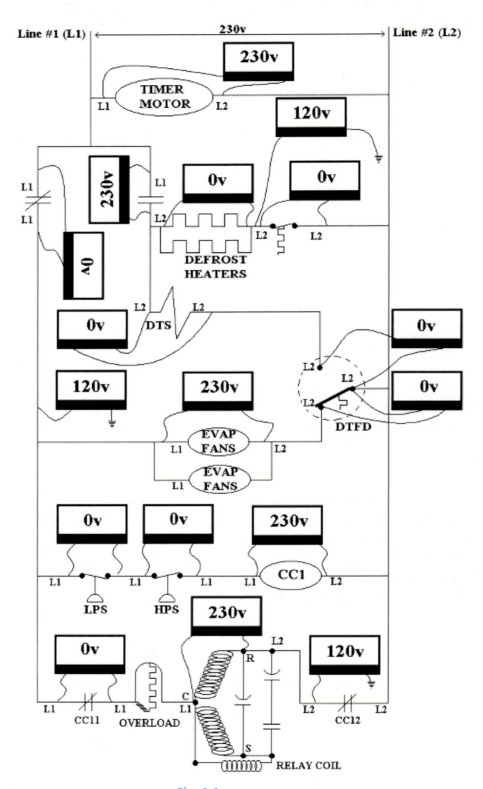

Fig. 6-3

OHMMETERS

Ohmmeters are used for two primary purposes:

1. To measure a circuit for continuity (a complete path for electron flow)
2. To measure the amount of resistance in a circuit or component

When measuring for continuity, the circuit is said to be open (infinity measurement) or completed or closed (no resistance). Fig. 6-4

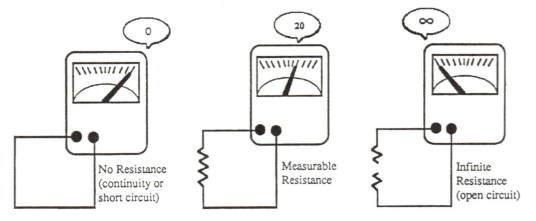

Fig. 6-4

An ohmmeter is only used on a circuit that has been disconnected from the power source. The ohmmeter has its own power supply (battery) to make resistive measurements.

 Most ohmmeters are protected by a fuse, but connecting an ohmmeter to a live circuit can still damage the meter.

When testing a circuit for continuity, test leads are connected in much the same way as the voltmeter leads in the example below. When testing from L1 to N (with the power off) the meter should show continuity or low resistance through all closed controls and safeties. Motor windings can be tested individually for the correct resistance value. Disconnect the component being checked from the circuit to prevent feedback from parallel circuits that can cause a false reading. Check manufacturer specifications for the resistance of any power consuming devices. Fig. 6-5

HP	Run	Start
1/8	4.5 Ω	16 Ω
1/6	4.0 Ω	16 Ω
1/5	2.5 Ω	13 Ω
1/4	2.0 Ω	17 Ω

Fig. 6-5

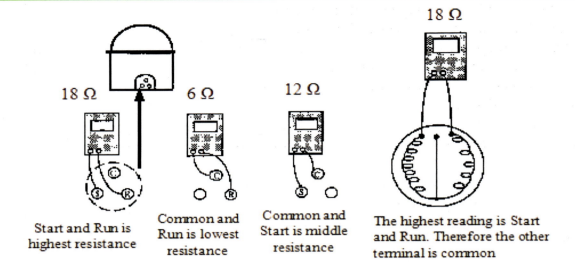

Fig. 6-6: Three ohmmeter readings are required to locate the common terminal.

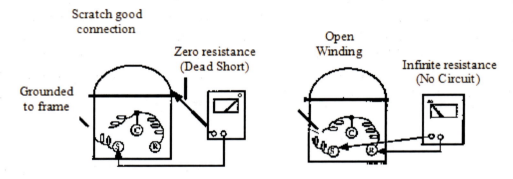

Fig. 6-7: Using an ohmmeter to check or "ring out" motor windings.

AMMETERS

Ammeters are used to measure the actual amount of current flowing in a circuit. This allows us to determine whether a power consuming device is actually operating. The example in Figure 6-8 shows the heating elements in an electric furnace. The elements are sequenced so that they are energized one at a time. By connecting a clamp-on ammeter, a technician can see the current increase as each element is energized.

Since each element has a rating of 5,000 watts, the amp draw of each element can be calculated using Ohm's law.

   I = P / E
   I = 5000 / 240
   I = 20.8 amperes

When the circuit is energized, the first element should measure approximately 20.8 amps. As the sequence timer energizes the second element, the reading should increase to approximately 42 amps. When the third element is energized, the current should increase to approximately 63 amps.

A voltmeter can be used to check the voltage applied to each element, but will not indicate whether the element is functioning.

An Ohmmeter can be used, with the power disconnected, to check the resistance of each element to ensure that none are open, but this is very time consuming.

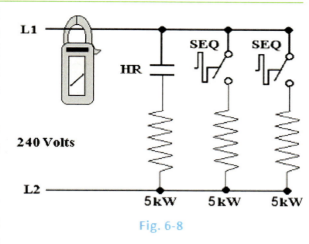

Fig. 6-8

An ammeter check is a much quicker way of finding out if each element is operating. It shows us whether there is voltage to the load, current flow, and a complete circuit.

When measuring small amounts of current, a clamp-on ammeter may lose accuracy. To increase the accuracy of the meter, a coil of wire can be used to increase the measured amperage. Coil a piece of wire into exactly 10 loops and jumper the coil into the circuit, in series with the load. Clamp the ammeter through the coil. The meter will now read ten times the amount of current in the circuit. For example, if the meter is showing 10 amps, there is 1 amp of current flow in the circuit. The measured amperage is divided by the number of loops or coils of wire around the jaws of the clamp-on ammeter.

The following four articles entitled "Troubleshooting Switches," "Troubleshooting Using a Voltmeter," "Systematic Troubleshooting," and "Voltmeter or Ohmmeter?" are written by Professor John Tomczyk of Ferris State University. Professor Tomczyk is the author of "Troubleshooting and Servicing Modern Air Conditioning and Refrigeration" and the co-author of some editions of "Refrigeration and Air Conditioning Technology."

TROUBLESHOOTING SWITCHES

Service technicians often encounter electrical problems when troubleshooting HVACR equipment. These problems are often nothing but electrical switches that are either opened or closed. However, power-consuming devices (electrical loads) are often in

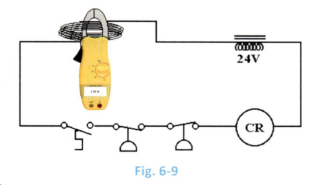

Fig. 6-9

series with these switches and can complicate matters. Most of the time, electrical switches are in series with one another and are relatively simple to troubleshoot electrically. It is when these electrical switches are in parallel with one another that it gets a bit more complicated.

Figure 6-10 on the following page shows a switch in series with a motor (electrical load or power consuming device). In this case, the electrical load is a Permanent Split Capacitance (PSC) motor. The potential difference (voltage) between Line 1 and Line 2 is 230 volts. This means that if the two leads

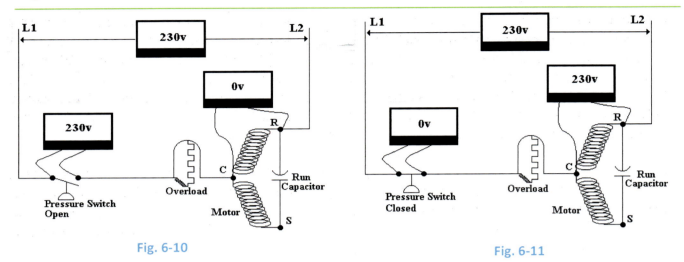

Fig. 6-10   Fig. 6-11

of a voltmeter were placed between Line 1 and Line 2, the meter would read 230 volts. If a technician measured the voltage across the open switch, a voltage of 230 volts would also be read on the voltmeter. This happens because Line 1 ends at the left side of the open switch, and Line 2 simply bleeds through the run winding of the motor, through the closed overload, and ends at the right side of the switch. Since the motor is not running or consuming power, then the windings are nothing but conductive wire for Line 2 to bleed through. If a technician were to measure the voltage across the run winding (between R and C) of the PSC motor, the voltage would be 0 volts because the motor winding is simply passing Line 2 when it is not running. Line 2 would be at both the R and C terminals of the motor, and the potential difference or voltage difference between Line 2 and Line 2 is 0 volts. However, if the technician references either the R or C terminal to the ground, the voltage would be 115 volts.

In Figure 6-11, the switch is closed and the motor is running. The motor is now consuming power and will drop the entire line voltage of 230 volts across its run winding (terminals R and C) while it is running. If the technician measures voltage across terminals R and C of the motor, 230 volts will be measured. A voltage measurement across the closed switch will read 0 volts. This is because Line 1 ends at the C terminal of the motor it is running. The switch experiences Line 1 at both its terminals, and the potential difference or voltage difference between Line 1 and Line 1 is 0 volts. If a technician measures from one terminal of the switch to the ground, the voltage reading will be 115 volts.

In the previous examples with switches and motors, the voltmeter across the open switch reads 230 volts, while the closed switch reads 0 volts. If one concluded that open switches always read voltage and closed switches always read 0 voltage, that would be wrong.

The next example will clarify this concept.

Figure 6-12 shows two switches in parallel, but at the same time in series with a motor (power consuming device). The top switch is closed,

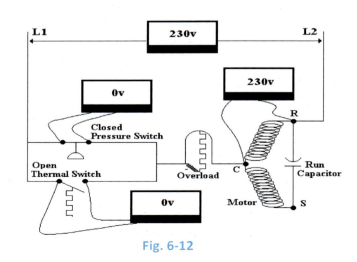

Fig. 6-12

but the bottom switch is open. A voltmeter across terminals A and B of the top switch will read 0 volts because it is measuring a potential difference between Line 1 and Line 1. Since the motor is running, it is dropping 230 volts across the run winding (C and R terminals) of the PSC motor. However, a voltmeter across terminals C and D of the bottom switch will also read 0 volts. This happens because Line 1 actually extends to the common terminal of the motor when it is running. This would make terminals C and D of the bottom switch both Line 1, and the voltage difference between Line 1 and Line 1 is 0 volts. Actually, points A, B, C, and D are all Line 1.

Figure 6-12 is a scenario where both an open switch and a closed switch read 0 volts. What technicians must do when electrical troubleshooting is to ask themselves where Line 1 and Line 2 are, not whether the switch is opened or closed. Measuring across the same line will always give 0 volts. Measuring across Line 1 to Line 2 will always give the total circuit voltage, which in these examples was 230 volts.

SYSTEMATIC TROUBLESHOOTING

Many service technicians are faced with troubleshooting refrigeration and air conditioning systems on a daily basis. Learning the most efficient methods of troubleshooting saves the service technician time as well as the customer money. Good systematic troubleshooting techniques are a win-win situation for both customer and service technician. Following is an example of a systematic troubleshooting method incorporating a symptoms/cause method.

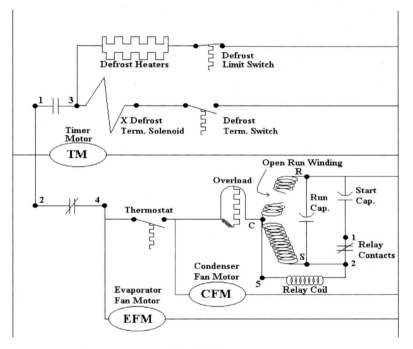

Fig. 6-13

Figure 6-13 illustrates an electrical schematic diagram showing a time clock controlling a defrost circuit and a refrigeration circuit. Notice that in the refrigeration circuit, the compressor's run winding has been opened by a motor overheating problem. The service call is a "no cooling" call for a low-temperature walk-in cooler. Once the technician quickly looks the system over and listens for any clues of what the problem may be, the electrical schematic, if available, should be studied. Understanding the logic or sequencing of the circuits before diving head-over-heals into the problem is of utmost importance in systematic troubleshooting.

In this scenario, an open run winding will cause certain symptoms not caused by other possible system problems. For example, the technician listened to and examined the refrigeration system and then studied the electrical schematic drawing. The service technician then lists the symptoms:
1. Compressor motor hums and will not turn
2. Compressor motor draws locked rotor amps (LRA)

3. Compressor motor's overload trips soon after drawing LRA
4. Resets after two minutes

The service technician then turns off power to the refrigeration system to let the motor cool down. After studying the electrical diagram again, the technician lists some of the possible causes that correlate to every symptom listed. If a possible cause does not correlate to *every* symptom, it cannot be a possible cause.

Possible Causes:
1. Open start winding
2. Open run winding
3. Open run capacitor
4. Open start capacitor
5. Compressor mechanically stuck
6. Potential relay contacts between 1 and 2 stuck open

Notice that every possible cause listed correlated to all the symptoms. Now all the service technician has to do is to check the six possible causes to find out which one is causing the symptoms, instead of blindly checking the entire system.

With the power off, the technician takes a wire off of the start winding and measures the windings resistance. The technician finds that it has 4 ohms, meaning that it is not open. The technician now takes a wire off the run winding and measures the windings resistance. The technician finds that it is an open winding because of the infinity reading on the ohmmeter. The compressor has to be replaced. With either winding open, the compressor has no phase shift for starting and will lock its rotor drawing LRA until the overload trips.

If either capacitor was bad, the motor may not have had enough phase shift to start. In certain cases, the motor may slowly turn. If the compressor was mechanically stuck, such as something wedged between the piston and cylinder, the motor would lock its rotor and draw LRA. If the contacts between terminals 1 and 2 of the potential relay were stuck open for some reason, the start capacitor would be out of the circuit. This again will probably not cause enough phase shift to start the motor turning. The motor would again draw LRA. Notice that in every case, all the symptoms were met.

What about an open overload or an open potential relay coil between terminals 2 and 5 of the potential relay? If the overload were opened at the beginning, maybe from too high of a compressor amp draw, the compressor would not hum or draw LRA. This would not correlate with all the symptoms listed and could not be a possible cause. If the coil of the potential relay were open, the contacts between 1 and 2 of the potential relay would stay in their normally closed position and not open. This would cause the start capacitor to be in the circuit all the time, and the motor would turn, draw higher than normal amperage, eventually open the overload and probably burn out the start capacitor. These symptoms do not correlate with the original symptoms listed, thus cannot be a possible cause.

Once the service technician has replaced the compressor and the system is up and running, it is important to run a system check to see what caused the compressor overheating that opened the winding. Evaporator superheat, total superheat, and condenser sub-cooling, along with suction pressure and head pressure, must be taken for the system check. In this case, the technician took a system check and found the evaporator superheat to be very high at 40° and the total superheat to be very high at 90°. Condenser sub-cooling was fine at 12°. Both suction and head pressure were low (see Figure 6-14). The problem that caused the overheating was a faulty thermostatic expansion valve. The valve would not open enough, and the entire low side of the system was being starved. The compressor was a refrigerant-cooled compressor. This caused the compressor to overheat and cycle on its overload until the run winding finally opened. Without this final system check, the new compressor would surely fail in a short time.

Table 1. R-134a	
Evaporator Superheat	40°
Total Superheat	90°
Condenser Subcooling	12°
Suction Pressure	4" Hg vacuum
Head Pressure	90 psig

Fig. 6-14

VOLTMETER OR OHMMETER?

Often service technicians will encounter switches in series or parallel with electrical loads. Keeping the electrical power on and using a voltmeter to voltage troubleshoot is the fastest and most reliable method. However, there will be times when a technician must switch to an ohmmeter and shut the electrical power off in order to get to the root of the problem.

Figure 6-15 shows an electric PSC motor in series with two switches that are in parallel with one another. The voltage between points A and B (the open switch), in this case, would be zero volts because the voltmeter would be measuring between Line 1 and Line 1. The voltage between points C and D (the closed switch) would also be zero volts because of the voltmeter measuring between Line 1 and Line 1 again. Remember, the motor is running and dropping all of the 230 volts while it is consuming power. A voltmeter across the R and S terminals of the PSC motor would read 230 volts because the meter is measuring the voltage between Line 1 and Line 2, which is 230 volts.

Figure 6-15 showed us that a voltmeter across the R and C terminals of the PSC motor would read 230 volts because it is measuring between Line 1 and Line 2. However, what would the voltage be between R and C if the run winding between R and C opened, causing the motor to stall and draw

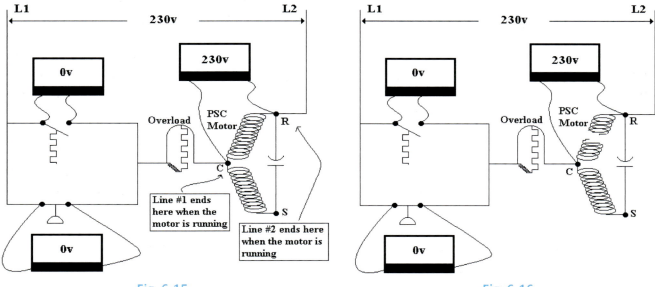

Fig. 6-15   Fig. 6-16

Locked Rotor Amps (LRA)? This is before the overload had opened. Figure 6-15 illustrates this scenario. Notice that a voltmeter placed across the R -and C terminals of the motor (the opened winding) will again read 230 volts. In fact, all the voltages in Figure 6-15 and Figure 6-16 are the same. Figure 6-16 illustrates that whether the motor is running properly or if it has an open winding, the voltage will still read 230 volts across R and C. So how does the service technician determine if the run winding is open or not? The answer is with an ohmmeter.

The service technician must shut down and disconnect one wire, either from the R or C terminal of the motor (Figure 6-17). Disconnecting the wire will prevent electrical parallel feedback from the ohmmeter's internal voltage source through another parallel electrical circuit. The technician must then place an ohmmeter across the R and C terminals of the motor. The measurement will read "infinite ohms" if the winding is open. This is the only way the service technician can tell if the winding is open or not.

Figure 6-18 shows a parallel feedback circuit from the ohmmeter's internal voltage source if a wire was not disconnected from the motor terminals. In this case, the ohmmeter reading would be 2 ohms. This could fool the technician into thinking the winding was still good.

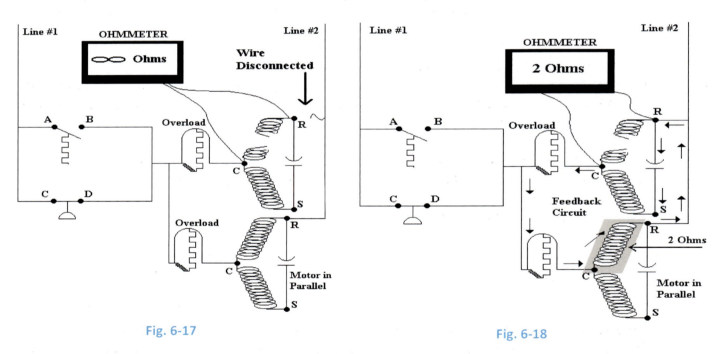

Fig. 6-17   Fig. 6-18

TESTING CAPACITORS

Capacitors (run or start) can be tested and the actual capacitance found using a capacitor tester found on most digital meters. An ohmmeter can be used to check basic conditions of a capacitor, such as a shorted or open circuit and whether or not the plates will take a charge.

 When testing a capacitor with an ohmmeter, the capacitor must first be discharged in order to prevent damage to the capacitor or ohm meter. Do not short across the terminals with a screwdriver; the proper method is to use a 15,000 ohm, two watt resistor connected across the capacitor terminals. This will allow the current to dissipate slowly and avoid damaging the capacitor.

Some capacitors are fitted with a bleed resistor. The resistor bleeds off the charge each time the circuit is de-energized. The bleed resistor must be carefully removed before continuing with testing.

This ohmmeter test is valid in diagnosing open or shorted capacitors; these are definitely bad. However, because a capacitor passes the ohmmeter test, you cannot guarantee the capacitance rating.

When checking a dual run capacitor, connect one meter lead to the center terminal (common) and the other lead to either herm. or fan and analyze the results for each. The leads between either terminal and the metal case of a run capacitor should show an infinite reading.

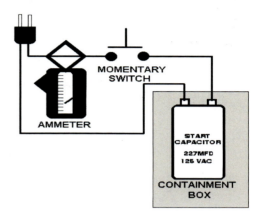

Once the capacitor has been safely discharged, connect the leads to each terminal of the capacitor allowing time for the meter to analyze the results. Faulty capacitors can be open, shorted, or have improper microfarad (mfd) ratings. A good run capacitor rating should be the same as indicated on the label.

Fig. 6-19: Testing a start capacitor

To maintain motor efficiency, always replace a weak run capacitor. Start capacitors usually have a range for the rating. Some motor manufacturers recommend a replacement start capacitor be within plus or minus 10 percent of the listed rating. Always use what is recommended by the manufacturer.

If you do not have a capacitor tester, you can measure and calculate the capacitance in microfarads (µf) by using an ammeter, a voltmeter, and a test cord, as shown in Figure 6-19.

Calculate the capacity by using the formula:

$$\text{Capacitance } \mu f = \frac{2650 \times \text{Amps}}{\text{Applied Voltage}}$$

Place the start capacitor in a containment box before energizing. In the event of a failure, a start capacitor can overheat and expel flaming material. Before connecting the test cord, perform the ohmmeter test. If the capacitor is shorted, do not proceed with this test!

Before connecting power to the capacitor, measure the voltage supply used for the test and record the measurement to be used in the calculation.

Close the momentary switch for three seconds and record the current flowing into the capacitor.

 IMPORTANT: Do not energize a start capacitor for more than three seconds.

If the motor's **start capacitor** capacitance is within plus or minus 20 percent of the low and high rating on the label, it is safe to use. If capacitance is below the label rating, the capacitor is breaking down and must be replaced.

Run capacitors are tested using the same procedures described for start capacitors, with a few exceptions. A run capacitor is housed in a metal case, therefore, when ohmmeter testing, check that there is no continuity from either terminal to the case. Because a run capacitor is designed for continuous duty, there is no time limit for energizing the capacitor. A **run capacitor** must be within ±10 percent of the rating.

TIPS AND SUGGESTIONS

Sometimes a PSC fan motor will have a "dead spot" condition. When the motor stops and the rotor aligns in just the right place, the motor will not start or run. If the rotor is turned left or right from its position, it will start and run, but will stop again and cause the same problem sometime in the future. Replace the motor with a new one.

A three-phase motor with a blown fuse in one of the three legs will try to run on the remaining two legs. This is referred to as "single-phasing."

Always check airflow and the rotation of the fan motor to insure that the blade is turning in the right direction. Sometimes a motor has been installed incorrectly.

If a fan motor shaft is hard to turn by hand, usually the problem is dry or seized bearings. On residential units, replace the fan motor. Larger, more expensive motors can sometimes be taken to a motor shop for repairs.

MOTOR TROUBLESHOOTING FLOW CHART

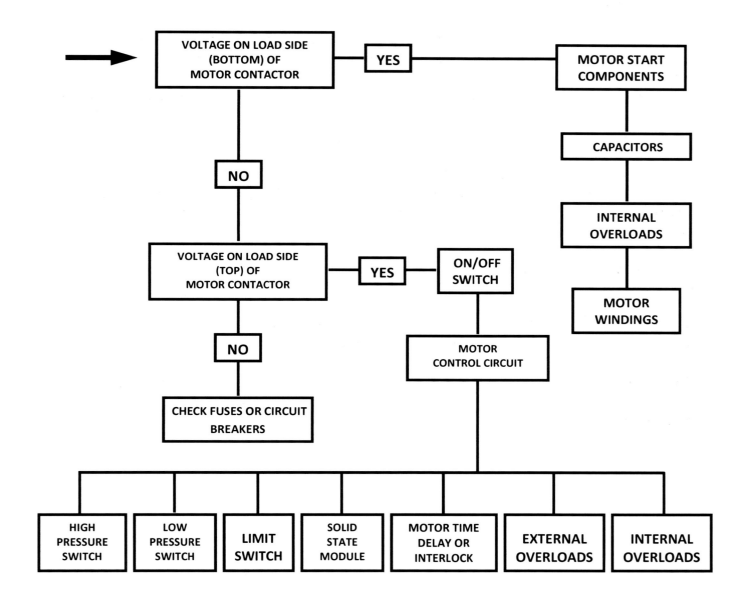

Student Worksheet

Chapter 6: Troubleshooting

Name _____   Date _____

1. When troubleshooting most HVAC-R loads, a technician is most likely to find loads wired in _____ with other loads.
    - A. series
    - B. parallel
    - C. complex detail
    - D. sequence

2. When troubleshooting switches, a technician will likely find them wired in _____ to a load.
    - A. series
    - B. parallel
    - C. complex
    - D. order

3. Power consuming devices:
    - A. are always high-voltage
    - B. are low-voltage only
    - C. can be a range of voltages
    - D. are mechanical motion only

4. A technician finds that a unit's indoor fan is running, but the condensing unit's fan and compressor are not running. Which of the following devices can be eliminated?
    - A. thermostat
    - B. transformer
    - C. fuse or breaker
    - D. condenser contactor

5. A technician is checking an A/C unit and finds that the indoor fan is not running, but the condenser unit is running. The problem is in the thermostat. Which terminal should be checked first?
    - A. R
    - B. Y
    - C. W
    - D. G

6. The load side of a switch is:
    - A. the side the power side is connected to
    - B. the side a power using device is connected to
    - C. connected to white wire only
    - D. the neutral side of the circuit

7. Which of the following service calls or troubleshooting problems are the most common for technicians?
    A. electrical problems
    B. refrigerant choices
    C. thermostat in the wrong place
    D. dirty condenser

8. Making unsystematic measurements or randomly replacing parts:
    A. rarely solves problems
    B. works best with most problems
    C. cannot ever repair the unit
    D. is the prescribed systematic method

9. Which of the following is an operational control?
    A. contactor
    B. thermostat
    C. fan limit switch
    D. overload

10. Voltmeter leads are placed across the switch for a burning light bulb operating on 120 volts. How many volts AC should it measure?
    A. half of the applied voltage
    B. all of the applied voltage (120 volts)
    C. no voltage across the control
    D. the voltage from the bulb or load

11. Using an ammeter to measure the total amperage of three 5-kW, 240-volt, single-phase heaters, how many amps should be measured after all three elements are energized and sequenced on?
    A. 15 amps
    B. 16 amps
    C. 62.5 amps
    D. 87.5 amps

12. What method is best for electrical troubleshooting on heating/cooling equipment?
    A. the methodical procedure for troubleshooting: understanding Ohm's Law, rules for series and parallel circuits, and related components.
    B. learning everything from observing someone on the job
    C. the practical method of changing the parts out until it works
    D. always calling someone to do it for you until you think you may know how

13. What diagram for electrical wiring has loads/components in parallel across lines 1 and 2?
    A. simple series
    B. pictorial
    C. ladder
    D. symbols

14. Pictorial diagrams have:
    A. pictures and a ladder
    B. all pictures and no wiring
    C. pictures or symbols with wiring colors and locations
    D. wire sizes without symbols

15. Wiring diagrams always show the unit with the power;
    A. "off," unless the instructions located on the diagram say differently
    B. "on" continuously, with nearly all air conditioning units
    C. partially on
    D. timed "off for several minutes"

16. A "dead spot" on a permanent split-phase motor is:
    A. a place in the stator where a faulty rotor sometimes lines up and will not allow the motor to start
    B. when the stator is locked by higher or lower voltage
    C. caused by dirt and build-up from over-lubricating the bearings
    D. when the motor is not operating as a PSC, but as a CSR motor

17. Which of the following meters is used with a 10-wrap multiplier?
    A. volt meter
    B. ohmmeter
    C. clamp-on ammeter
    D. line finder meter

18. While checking a 120-volt simple circuit containing a fuse, switch, and small shaded pole motor, the volt meter measures 118 volts across the motor, but the motor is not running. The technician concludes that the _____ is defective.
    A. motor
    B. fuse
    C. switch
    D. wiring

19. When checking a condensing unit contactor, a technician finds 115 volts across the contacts when the coil is energized. The technician should:
    A. assume the contactor is ok
    B. replace the contactor
    C. clean the contacts with small file
    D. clean the contacts with solvent cleaner

20. A condensing unit cycles on and off about every 5 to 8 seconds. Using a volt meter while the unit is cycled off, a technician finds 24 volts across the low-pressure switch and 0 volts across the high-pressure switch. The diagnosis should be:
    A. unit is cycling because of low pressure
    B. unit is cycling because of high pressure
    C. there is no problem with either switch
    D. the problem is not electrical

Glossary

A

alternating current (AC)
electrical voltage and current that changes in magnitude and direction in a cyclic characteristic; AC electricity in North America cycles 60 times per second

ammeter
an instrument used to measure current flow through a circuit

ampere
a unit of current flow equal to one coulomb of charge moving through a circuit in one second

analog meter
a meter that displays information using a needle moving over a scale

anode positive terminal of an electrical device

atom
the smallest unit of an element, made up of protons, electrons and neutrons, (except hydrogen)

B

British thermal unit (Btu)
a unit of energy based on the amount of heat required to raise the temperature of one pound of water one degree Fahrenheit

C

capacitance
a measure of the energy stored in a capacitor; units of capacitance are farads

capacitor
a device made with two conductive plates separated by a dielectric used to increase an inductive load efficiency by changing the phase angle and electrostatically stores electrical energy

capacitor-start motor
a modification of a split-phase motor that uses a capacitor in series with the start winding; the capacitor causes a phase displacement for starting

capacitor-start induction-run motor
a motor in which the capacitor assists start under high load conditions; once started, the motor operates on the run winding only

capacitor-start capacitor-run motor
a motor that utilizes a start capacitor for high starting torque and a run capacitor for running efficiency

cathode
the negative terminal of an electrical device

centrifugal switch
a switch that uses a combination of weights and springs to open its contacts at approximately 75 percent of a motor's RPM; wired in series with the start winding and/or start capacitor

circuit
electrical wires and components that allow current flow in a complete loop, away from the energy source and back again

circuit breaker
a safety device designed to automatically open a circuit at a predetermined overload of current

clamp-on ammeter
current measuring device that senses the strength of the magnetic field around a conductor

coil
wire arranged in a spiral shape (usually around an iron core) that creates a strong magnetic field when current passes through it

common
a terminal, connection, or other part of an electrical circuit that is shared by different components

conduit
a tube used to carry and protect electrical wires; can be metallic or non-metallic

connection
mechanical or electrical joining of two parts

contact points
movable points that complete a circuit when pressed together; usually made of tungsten, platinum, or silver

continuity
an unbroken line or continuous path through which electricity can flow

control
an automatic or manual device that directly operates electrical supply to the equipment

coulomb
the quantity of electricity equal to a current of one ampere in one second; one coulomb equals 6.25 x 10^{18} electrons passing a point in one second

counter electromotive force
action that takes place in motors when voltage (EMF) is self-induced in the rotor conductors which is opposite of the source voltage

covalent bond
atoms joined together to form a stable molecule by sharing electrons

current
movement of electrons in a conductor, usually expressed in amperes; symbol is I (for intensity)

current relay
a relay operated by the starting current of a motor; allows the start winding to drop out of the circuit after the motor starts

cycle
the voltage generated by the rotation of a conductor through a magnetic field, from a zero reference in a positive direction, back to zero in a negative direction; one complete cycle equals 360 degrees of rotation

cycles per second
the number of alternating current waves in one second of current flow

D

data plate
equipment identification label, usually containing information such as model, serial number, voltage, and amperage

dead
describes a portion of a circuit with no voltage

dead leg
the grounded phase of a three-phase delta wound transformer

dead short
a very low-resistance connection that allows the unrestricted flow of electrons

dedicated circuit
a circuit that is fused and supplies power to one appliance only

de-energize
to stop electron flow to a device or open a circuit

delta transformer
a three-phase transformer that has the finished end of one winding connected to the finished end of a second winding; configuration resembles the Greek letter delta (Δ)

diac
a semiconductor most often used as a voltage-sensitive switching device

dielectric
an insulating material separating the conducting surfaces of a capacitor

digital voltmeter
a voltmeter that uses direct numerical display as opposed to a meter movement

diode
a solid-state device that allows current to flow only in one direction; will rectify alternating current to direct current

direct current (DC)
current that flows in only one direction in a circuit

distribution center
an electrical panel that supplies electricity to several places in a structure

doping
adding an impurity to a semiconductor to produce a desired change in electrical properties

double-pole breaker
a circuit breaker used to disconnect both hot wires with a single on-off action

double-pole double-throw switch
a switch with two poles and two contacts for each pole; two contacts are always open and the other two always closed

double-pole single-throw switch
a switch arranged so that both switches are either open or closed

duty cycle
the relationship between operating time and off time; duty cycle of a motor is usually referred to as continuous or intermittent

E

E
symbol for voltage (electromotive force)

earth
term for zero reference ground

eddy current
induced current flowing in a magnetic core created by a varying magnetic field

Edison base fuses
15, 20, and 30 amp fuses with the same style of base as incandescent bulbs

effective voltage
a value of an AC sine wave voltage with the same heating effect as an equal value of DC voltage; $E_{eff} = E_{peak} \times 0.707$

electrical charge
a basic property of elementary particles of matter; a charge that can be either positive or negative; like charges repel and unlike charges attract

electrical degree
one 360th of an alternating current or voltage cycle

electricity
energy produced through the flow of electrons from one atom to another

electromagnet
a magnet made by passing current through a coil of wire wound around an iron core

electron
the part of an atom that carries a negative charge

ELI
an acronym used to remember that voltage (E) in an inductive (L) circuit leads current (I) or current lags voltage

EMF
abbreviation for electromotive force

EMI
abbreviation for electromagnetic interference

EMT
abbreviation for electrical metallic tubing (thin-wall conduit)

end bell
the plate at the end of a motor that supports the bearings; also called an end shield or end plate

energize
to supply power or electron flow to an electrical circuit

Energy Efficiency Ratio (EER)
the number of Btu produced per watt of electrical power consumed by an air conditioner

Environmental Protection Agency (EPA)
a U.S. government agency dedicated to writing and enforcing environmental and human health regulations

F

F
abbreviation for farad, frequency, fluorine, or Fahrenheit

factual diagram
a wiring diagram that is a combination of pictorial and schematic diagrams. A schematic with component locations.

farad
a unit of electrical capacity; capacity of a device that gives a difference of one volt of potential when charged with one coulomb of electricity

fast-acting fuse
a fuse that opens quickly on overloads and short circuits; not designed for temporary overloads that occur with inductive or capacitive loads

field
an electrical or magnetic area of force

field pole
the part of a stator that concentrates the magnetic field of the field winding

fish tape
flexible wire used to pull wires through conduit

forward bias
voltage applied to a P-N junction diode to neutralize the potential barrier (positive voltage on a P-region or negative on an N-region)

fractional horsepower
a horsepower value less than one

free electrons
electrons in the outer orbit of an atom that are easily removed and result in electrical current flow

frequency
number of cycles that an AC current completes in one second, expressed in hertz (Hz)

full load amps (FLA)
current drawn by a motor when operating at rated load, voltage, and frequency

fuse (fu)
an electrical safety device consisting of a metal strip that melts when subjected to high current

G

generator
a mechanical rotating electric device that converts mechanical energy to Direct Current (DC) electrical energy

Greenfield
a flexible metal conduit used in applications that require bends at various angles

ground (GRD)
n. a common point of zero potential such as a motor frame or chassis
v. to connect a circuit to the earth, thus making a complete circuit

ground fault interrupter circuit (GFIC or GFI)
an electrical device that opens a circuit that has been accidentally grounded

H

hard-start kit
a kit consisting of a start relay and capacitor to provide high starting torque

hertz (Hz)
a unit of frequency equal to one cycle per second

high-voltage circuit
a circuit involving a potential of more than 600 volts

horsepower (HP)
measure of the amount of work a motor can produce during a period of time

hot wire
a lead with a voltage difference between it and another hot wire or a neutral wire

HVACR
heating, ventilating, air conditioning, and refrigeration

I

I^2R
formula for finding power in watts

ICE
an acronym used to help remember that current (I) in a capacitive (C) circuit leads voltage (E)

impedance
total resistance to electron flow in an AC circuit caused by resistance, inductive, and/or capacitance reactance measured in Ohms, represented by the symbol Z

induced current
current that flows as the result of an induced voltage

induced voltage
potential that causes current to flow in a conductor that passes through a magnetic field

inductance
property of a circuit that allows energy to be stored in a magnetic field

inductive circuit
any circuit that contains at least one magnetic field

insulation
materials with few free electrons used to cover wires to prevent short circuits and shock hazards

integrated circuit
a circuit with multiple semiconductors and transistors in a single circuit; sometimes called a chip

ion
an atom that has gained or lost electrons, resulting in a positive or negative charge

J

jacket
the outside cover of a wire or cable

junction box (j box)
a plastic or metal box where electrical connections are made

jogging
quickly and repeatedly switching a motor on and off

joule
the amount of heat needed to raise the temperature of one kilogram of water 1°C; equal to one watt second

jumper
a wire placed across the contacts of a component for test purposes

junction
a point in an electrical circuit where the current branches out into other sections

K

Kirchoff's Current Law
the sum of the currents flowing into any point or junction of a circuit is equal to the sum of the currents flowing away from that point

Kirchoff's Voltage Law
in any current loop of a circuit, the sum of the voltage drops is equal to the voltage supplied to that loop

kV
abbreviation for kilovolt

kVA
abbreviation for kilovolt amps

kVAh
abbreviation for kilovolt amp hour

kW
abbreviation for kilowatt

kWh
abbreviation for kilowatt hour

kilo-
prefix meaning one thousand

L

L
symbol for inductance measured in Henrys

L1
incoming power supply line

L1-L2
incoming power supply lines

L1, L2, L3
incoming three-phase power supply lines

ladder diagram
a circuit diagram drawn in the form of a vertical ladder

Law of Magnetism
unlike poles of a magnet attract and like poles repel

Lenz's Law
the induced counter electromotive force in a circuit will always be in a direction that opposes the voltage that produces it

line
conductors carrying power from the generating source

load center
point from which branch circuits originate

locked rotor amps (LRA)
steady state current drawn by a motor when the rotor is locked to prevent its movement

lockout (safety)
the opening and locking of the main power switch to safely perform service procedures

lugs
terminals on the end of a wire or places on equipment to facilitate connections

magnetic field
lines of force that extend from a north polarity and return to a south polarity to form a loop around a magnet

magnetic flux
lines of force that connect the north and south poles of a magnet

main
the primary circuit supplying all other circuits

make
to complete a circuit by closing a switch

meg- (or mega-)
prefix meaning one million

mho
unit of electrical conductivity of a body with a resistance of one ohm (reciprocal of ohm)

megohm
unit of resistance equal to one million ohms

micro-
prefix meaning one millionth

microfarad (mfd)
one millionth of a farad

milli–
prefix meaning one thousandth

milliampere (mA)
one thousandth of an ampere

millivolt (mV)
one thousandth of a volt

momentary switch
a spring loaded control that makes or breaks a circuit only when it is held in place

motor
a device that changes electrical energy to mechanical motion

multimeter
a meter capable of two or more electrical quantities, such as volts, amps, or ohms

N

National Electrical Code (NEC)
a national code written for the purpose of safeguarding persons and property, sponsored by the National Fire Protection Association

negative charge
the charge that results from an excess of electrons

negative temperature coefficient thermistor (NTC)
a resistor that decreases resistance as temperature increases

neutral
having no charge (line which is grounded at the fuse box)

neutron
a particle with no electrical charge located in the nucleus of an atom

nominal
the average rating of power or voltage during normal operation

nonferrous
the group of metals that do not contain iron

normally closed
describes a device that automatically moves to a closed position when power is removed

normally open
describes a device that automatically moves to an open position when power is removed

O

ohm
unit of electrical resistance; symbol is Ω

ohmmeter
instrument used to measure resistance

Ohm's Law
voltage equals current times resistance

open circuit
an interrupted electrical circuit that does not provide a complete path for current flow

oscillator
a device that changes DC voltage into AC voltage

oscilloscope
a cathode ray tube (CRT) that displays voltage (vertical scale) and time (horizontal scale)

out-of-phase
the condition resulting when two components do not reach their positive and negative peaks at the same time

overcurrent
a condition in an electrical circuit when the normal current is exceeded; can be caused by an overload or a short circuit

P

pos
abbreviation for positive

panel box
electrical junction box that contains fuses or breakers

parallel circuit
a circuit that feeds identical voltage to all branch circuits or components, with amperage dividing among the components

peak load
the maximum load carried by a unit during a designated period of time

peak-to-peak voltage
the measurement of voltage from the positive peak to the negative peak of an AC sine wave

permanent split capacitor (PSC) motor
a single-phase motor that has a capacitor continuously in series with the start winding

phase angle
the degree of difference between two AC sine waves

pole
one set of contacts, such as in a switch or relay

polyphase generator
a generator that rotates three conducting loops (three-phase)

polyphase motor
a motor that operates on three-phase current

positive charge
the charge that results from an excess of protons or lack of electrons

positive temperature coefficient thermistor (PTC)
a thermistor that increases resistance as temperature increases

potential relay
a relay with normally closed contacts and a coil with very high resistance that is energized (opened) by counter EMF from the start windings

primary winding (pri)
the coil of a transformer to which source voltage is applied

proton
a positively charged particle located in the nucleus of an atom

Q

quick-connect
a solderless terminal with a push-on connection

R

rf
abbreviation for radio frequency

rpm
abbreviation for revolutions per minute

raceway
a channel, conduit, or runway for conductors or cables

reactance
the effects of capacitance and/or inductance on an alternating current circuit, expressed in ohms; symbol is X

rectification
the process of converting AC to DC

relay
an electromechanical device that can be energized with a relatively small current to operate a set of contacts carrying a larger current

resistance
opposition to current flow

rheostat
a variable resistor that can be adjusted to various levels

root mean square (rms)
the effective value of an alternating periodic current or voltage, calculated as the square root of the average of the squares of all instantaneous values of the current (or voltage) throughout one cycle

rotor
the rotating part of a motor or generator

run winding
the motor winding that carries current during normal operation

S

safety ground
the conductor that connects the equipment frame or chassis to earth ground; usually a green or bare wire

safety motor control
a device operated by pressure, temperature, or current that opens the circuit if safe conditions are exceeded

scale
a measurement band on a test instrument

secondary voltage
the output voltage of a transformer; can be higher or lower than primary voltage

series circuit
a circuit with only one path for current flow

service entrance
the point of entry from the main electrical power line into the building

short circuit
an unintentional connection of low resistance resulting in excessive and often damaging current flow between two points in a circuit

single-phase
describes a device that uses or produces only one alternating current

slip
the difference between the speed of the rotating magnetic field of a motor and the actual rotor speed

Soft start
the use of a Positive Temperature Thermistor (PTC) with or without a start capacitor connected in series with the motor start winding.

solenoid
an electromechanical device that moves an iron core when energized

start winding
a winding which operates out of phase to help in starting a single phase electric motor

switch
a device that makes or breaks contacts to either complete or open a circuit

T

tag-out
the practice of labeling switches to inform others that repairs are in progress

terminal
a connection point on an electrical device

three-phase current
a combination of three alternating currents that are 120 degrees different in phase

torque
twisting or rotating force measured in inch or foot pounds

transformer
a device with two or more electromagnetic coils, used to increase or decrease or isolate an AC voltage

transistor
a semiconductor used to perform switching or amplifying of electrical signals

U

Underwriters Laboratories (UL)
an independent testing agency for electrical appliances

utility transformer
a transformer that steps down the utility supply voltage for use in a facility

V

VAC
abbreviation for volts alternating current

valance electrons
the number of electrons in the outer orbit of the atom determine whether the material is a conductor, insulator or semiconductor

variable frequency drive (VFD)
an electronic device that varies frequency and voltage to control the speed of an AC motor

variable speed drive (VSD)
an electronic device that varies voltage through the use of pulse width modulation to control the speed of an AC or DC motor

VDC
abbreviation for volts direct current

VMAX
the maximum voltage in an AC cycle

volt
unit of electromotive force

voltage drop
the amount of voltage loss from the source through a conductor or load

voltage root mean square (VRMS)
the average voltage in a circuit; equal to peak voltage times 0.707

volt-ampere (VA)
unit of apparent electrical power

volt-ohm-milliammeter (VOM)
a meter with multiple functions and ranges, usually including voltage, current, and resistance

W

watt
unit of true electrical power; symbol is W

Watt's Law (Power law)
in a DC circuit or in a purely resistive AC circuit, watts equals volts time amps

waveform
a graphic display of voltage values over a period of time

wavelength
the distance between two corresponding points of two successive waves of a periodic waveform

wire
a conductor, bare or insulated

wire gauge
the system of wire sizing by diameter, #0000 for the largest, #40 and above for the smallest

wye connection
made by joining one end of each of three windings; also called a star connection

y-axis
the vertical axis on a graph

zener diode
a diode that has a constant voltage drop when operated in reverse direction; often used as a voltage regulator